The Story of
Peking Man

The Story of
Peking Man

Penny van Oosterzee

ALLEN&UNWIN

First published in 1999 by Allen & Unwin
Revised paperback edition published in Australia
2001 by Allen & Unwin
83 Alexander Street
Crows Nest NSW 2065 Australia
Phone: (61 2) 8425 0100
Fax: (61 2) 9906 2218
Email: frontdesk@allenandunwin.com
Web: http://www.allenandunwin.com

National Library of Australia
Cataloguing-in-Publication entry:

Van Oosterzee, Penny.
  The story of Peking man.
  [Rev. ed.]

  Bibliography.
  Includes index.

  ISBN 1 86508 623 0.

  1. Peking man. 2. Fossil hominids. I. Title.

569.9

Set in 11.75/16 pt Weiss by Bookhouse, Sydney
Printed by Brown Prior Anderson Pty Ltd, Burwood, Vic

10 9 8 7 6 5 4 3 2 1

Contents

For my mother

1

Dragon bones

The peasant heaved the long, smooth, wooden pole, shiny with use, into a comfortable position across the sinewy muscles of his shoulders. Two large bamboo baskets, bulging with lumps of limestone, hung from each end of the pole. The inscrutable face, shaded from the harsh northern Chinese sun by a conical bamboo hat, did not give any hint of his racing thoughts as he continued with his mundane activity, and merged with the lines of other look-alike peasants in loose smocks of coarse-woven cloth, seamless ankle-length trousers and bare feet, desultorily jogging down the dusty path, straining against their loads of limestone from the mine, south-west of Peking. The rock would be heated in the nearby kilns to supply the burgeoning building industry of the city.

The peasant's mind was not on his job. He wasn't thinking of any of the myriad mindless activities that comprised his days and nights. He was imagining dragons—Chinese dragons that breathed clouds, not fire, and enjoyed looping and diving in aerial displays that sent colours shimmering off their scales. As every Chinese knew, dragons enjoyed tossing giant pearls, like opalescent thunder balls, among the clouds, creating lightning—and when it suited them, rain. In an agricultural country, that meant life, and temples worshipping the life-givers were found throughout the length and breadth of China in any place of any importance. A sighting of an azure dragon heralded the onset of the spring thunderstorms and good rain for the coming agricultural season.

Divine creatures, Chinese dragons were beings that could shrink to the size of an individual silkworm or bulge to a size too large for the world to contain. They were shape-shifters and could create the head of a camel, the horns of a stag, the eyes of a demon, the ears of a cow, the neck of a snake, the belly of a clam, the scales of a carp, the claws of an eagle and the paws of a tiger. Or they could take the form of a human. They could soar above the clouds, or immerse themselves in the abyss of waters. Indeed, each sea, river or lake had its guardian dragon, kingly in status, living in a crystal underwater palace surrounded by priceless treasures. Rubies were fossilised dragon's blood. As guardian spirits, dragons belonged to the race of immortals and mixed freely with gods and goddesses, who sometimes used them as mounts as they rode about the sky.

On the ground, the landscape was criss-crossed with 'dragon-lines', veins of the earth through which natural forces pulsed, flowing along mountains, ridges and rivers. As every Chinese knew, it was important, before building a house or choosing a burial site, that a geomancer be consulted to divine the earth's surface, to ascertain whether the proposed development was likely to obstruct the natural forces flowing through the dragon-lines, thus arousing the dragon's anger and causing calamity.

Dragons were simply the most important of all mythical beasts found in the Chinese tradition, and although subject to rather short tempers, they were generally well-disposed toward humankind—they had even become the symbol for the perfect man, the son of heaven, the Emperor. Indeed the Emperor's throne was called the 'dragon's throne', and his robes were richly ornamented with embroidered dragons. The gold and silver dragon on the lapel of the imperial coat, mounted on black silk, was the imperial symbol of the Emperor. Early emperors could actually be sons of dragons. Indeed, the Emperor Yao, a legendary, proto-historical figure came into the world by the visit of a red dragon to his mother. Similarly the founder of the mighty Han dynasty, Kao Ti (who ruled from 206 to 195 BC) was born after his mother was visited by a dragon while she was asleep by the side of a lake in a storm. The woman dreamed that she was embraced by a god and later gave birth to Kao Ti.

Not only emperors were heralded by dragons. On the night when Confucius, the Great Teacher, was born (551 BC) two azure-coloured dragons descended from heaven to his mother's home.

The Chinese peasant never demanded a clear separation of the worlds of myth and reality. History and legend blended, and were at any rate so bound up that it was impossible to say where one began and the other ended; historical figures were made into dragon-gods and myths were recounted as history. The peasant himself was convinced his ancestry went back to the beginning of time, through the dynasties, moving seamlessly from an historical to a mythical time-scale.

The Great Beginning, otherwise known as Chaos, created the void, a great formless mist of such darkness, such unmoving and silent immensity that its origin cannot be known. This produced the Space–Time–Universe that formed the original Breath. A thing without limits, its clear and pure elements rose lightly to form the Sky; its heavy and gross elements coagulated to form the Earth. The complementary essences of Sky and Earth were the Yin and the Yang. The concentrated essences of the Yin and Yang formed the seasons. The warm breath of the Yang accumulated to form fire, whose subtle elements became the sun. The cold breath of the Yin accumulated to produce water. The most subtle of the superabundant solar and lunar elements became the stars. The Sky received the sun, moon and stars; the Earth received the waters, rivers, ground and the dust. The sky was a round canopy covering the square Earth and held apart from it by eight pillars, or mountains.

The symbols of the Yin and Yang were two divinities with the bodies of dragons. Their names were Fuxi and Nuwa. All powerful, their existence was not always without drama. When a horned monster Gonggong, the Black Dragon, made a hole in one of the pillars holding up the world with a thrust of his horns, the pillar tilted forcing the heavenly bodies to flow westward, while on earth rivers flowed towards the east. A great part of the mountain came tumbling down, a jagged hole was torn in the sky and great crevasses appeared in the earth. From these massive chasms, fire and water spewed forth, causing a great flood that covered the surface of the earth. Nuwa killed the dragon and mended the rent in the sky with the metals of earth.

Nuwa was also the creator of men, and in the deep past she began to intricately fashion men out of China's yellow earth. Initially she found the results pleasing but the task became tedious, so she drew some mud, which she twirled into myriad droplets, each droplet springing into the form of a man. The nobles were the men fashioned from yellow earth and the poor people, who lived in vile and servile conditions, were those drawn out of droplets of mud.

The peasant smiled at the ironic thought. No doubt he was poor and lived in vile and servile conditions. He was, after all, drawn from mud. But that day hadn't he been lucky? And hadn't he plucked from the ancient limestone—the hardened mud of ages—a precious tooth? He had found it jutting from a 'dragon' skull, which looked remarkably human and was itself embedded in a limestone chunk. Dragon bones were used in Chinese medicines and worth money if found in large accumulations. But in limestone they were not that common and at any rate were cemented into the rock, making easy extraction impossible. Teeth, however, were prized and even individually were valuable. He knocked out the jutting tooth with a small axe and secreted it among the folds of his smock because he knew he could sell it later to a Peking apothecary, keeping the returns for himself. He then smashed the rest of the limestone together with its treasure of bones for easy shipping in his baskets, and on camels to the lime kilns.

The peasant didn't give a second thought to the actuality of the fossil creature. It wasn't surprising to find dragon bones among the chunks of limestone since dragons had existed since the beginning of time.

There was no trouble finding an apothecary in the warren of shops opening onto the narrow streets of Peking since most apothecary shops had a large glass globe filled with coloured fluid, usually red, either suspended by chains at the entrance or hanging in the front window of the shop. These globes were the symbol of their profession and were a means of advertising the services available from within. Inside, the shop itself was a confusion of shelves with glass and ceramic containers filled with mostly unidentifiable substances. Glass containers were generally labelled with characters in gold, whereas ceramic jars had a paper label with red characters affixed to them.

It was long known to European amateur fossil collectors that fossils could be obtained from Chinese apothecaries and not only in China but also throughout South-East Asia, Europe, even

Australia—everywhere there were local 'China towns'. Notes in the European literature in the middle part of the nineteenth century had suggested—though no-one knew for sure—that 'dragon' bones were actually fossil remains of large animals, such as horses, rhinoceroses and elephants. To the Chinese, these obviously mighty animals must have been imbued with mighty medical powers. And this was worth money. Peasants therefore collected the fossilised bones where they were easy to extract and sold them to medicinal procurement centres and herbal medicine shops.

The most prolific remains of 'dragons' were found in geologically young landscapes consisting of plateaux of red clay and intersected by valleys and narrow ravines. Many of the 'dragon' remains occurred in veritable bone nests within the red clay walls of the ravines. Wherever an accumulation of bones was discernible, tunnels were dug into the ravine walls, often about 1 metre in height and breadth, big enough for a man to crawl through. When the supply of bones was exhausted, exploratory digging was continued until a new supply was found. In this way the ravine walls turned into a beehive of galleries, many over 100 metres in length, with industrious bone hunters working with simple pick axes deep into the clay.

The blocks of clay were conveyed from the mine galleries on small wooden trolleys running on four solid wooden wheels, and hauled by a man on all fours by a rope running from one shoulder between his legs to the trolley. Light was supplied by simple lamps burning local vegetable oil. The work, done in winter, alternated with the summer's agriculture. In the spring, the buyers for large apothecary firms arrived and the bones were sold, per pound, for small copper coins, while the much more valuable teeth were sold for large copper coins.

In millennium-old Chinese pharmacological works there was much information concerning dragon's bones and their medicinal properties. For instance, a thousand-year-old favourite recipe recommended that:

To use the dragon's bone first boil some aromatic herbs. Wash the bone twice in the hot water, then reduce it to powder and place it in bags of thin stuff. Take two young swallows and, after removing their entrails, stuff the bags into the swallows and hang them over a spring. After one night take the bags out of the swallows, remove the powder and mix it with a preparation for strengthening the kidneys. The effect of such a medicine is as if it were divine.

In later times the procedure seems to have been considerably simplified, for an author of the Sung period (960 to 1279 AD) writes that the bone should lie in spirit overnight, then be dried over a fire and then powdered. For a quick fix, the bone should merely be pulverised and the powder taken in tea.

Not all the sages, however, were in agreement about the relationship between dragons and dragons' bones. Some of them considered that the bones were remains of dead dragons, while others insisted that dragons changed, not only their skin, but also their bones. Of course one had to be careful about which dragon bones were used. Those which were narrow with broad veins were from female dragons; those which were coarse, with narrow veins, were from the opposite sex. Those showing five colours were best; the white and yellow were medium quality; and the black ones were worst. As a rule it was said that those longitudinal veins were impure, and those collected by women were useless.

According to the ancient pharmacopoeia, many diseases may be cured by dragon's bone: dysentery, gall-stones, fevers and convulsions in children at the breast, internal swellings, paralysis, women's diseases and malaria. Dragon's teeth were also highly esteemed as medicine and according to the oldest medical work, written by the mythological emperor Sheng Nung, dragons' teeth would drive out the following afflictions: spasms, epilepsy and madness, and the twelve kinds of convulsions in children.

According to another ancient author, dragons' teeth have the quality of appeasing unrest of the heart and calming the soul. According to a third, they cure headache, melancholy, fever, madness and attacks by demons. All the authorities were agreed on one point: dragons' teeth are an effective remedy for liver diseases.

In 1899, a German physician and amateur fossil collector, K.A. Haberer, travelled to China. Whether he visited the mysterious Chinese apothecaries' shops for his work, as a doctor, or his passion, as a palaeontologist, is unsure but, several thousand years after 'dragon' bones appeared in mythology, it was he who made the very first systematic collection of them.

Haberer must have spent most of his spare time in the mysterious, dim and stuffy atmosphere of the apothecaries sifting through broken pieces of 'dragon' bones and teeth. Under the circumstances, how he managed to do this is as marvellous as his discoveries.

Haberer had every intention of exploring the interior of China but due to events there, culminating in the Boxer Rebellion in 1900, he was forced to restrict his activities to what were known as the 'Treaty Ports' of Shanghai, Ningpo, Ichang and Peking.

Throughout the nineteenth century, China's emperors had watched as foreigners made inroads into their land. Time and again, foreigners forced China to make concessions, and foreign regiments armed with modern weapons consistently defeated entire imperial armies. At the turn of the century Tsu Hsi, Empress Dowager of the Ch'ing dynasty was fed up.

Austria, France, Germany, Great Britain, Italy, Japan and Russia were all claiming exclusive trading rights to various parts of China, essentially dividing the country as one might a dragon's treasure. The United States was also looking for a way in and, instead of fighting for exclusive trading rights to its own portion of China, it wanted the lot. The United States craftily suggested an 'Open Door' policy which would guarantee equal trading rights for all, and prevent one nation from discriminating against another.

While these eight powers quarrelled over China, Tsu Hsi issued an imperial message to all Chinese provinces:

> The present situation is becoming daily more difficult. The various Powers cast upon us looks of tiger-like voracity, hustling each other to be first to seize our innermost territories...Should the strong enemies become aggressive and press us to consent to things we can never accept, we have no alternative but to rely upon the justice of our cause...If our...hundreds of millions of inhabitants...would provide their loyalty to their emperor and love of their country, what is there to fear from any invader? Let us not think about making peace.

In northern Shantung province a devastating drought was pushing the inhabitants, who were already blaming foreigners for their impoverishment, to the edge of starvation. Here, there were no thoughts of peace, and a secret society, known as the Fists of Righteous Harmony, was attracting thousands of followers who were intent on destroying the foreigners who were flouting millennia-old Chinese traditions and disrupting family relations. Foreigners, lacking a Chinese imagination, called them Boxers because they practiced martial arts. It was said that The Fists of Righteous Harmony had magical powers and that foreign bullets could not harm them. Millions of 'spirit soldiers', too, would soon rise from the dead and join their cause.

In the early months of 1900, thousands of Boxers roamed the countryside. They attacked Christian missions, slaughtering both the foreign missionaries and their Chinese converts. Then they moved toward the cities, attracting more and more followers as they came. Nervous foreign ministers insisted that the Chinese Government stop the Boxers, but inside the Forbidden City, the Empress outwardly did nothing. Internally, however, she ordered that all foreigners be killed. Soon after, the German minister was the first to be murdered.

Foreign diplomats, their families and staff lived in a compound just outside the Forbidden City's walls in the heart of Peking. They hastily threw up defences, and with a small force of military personnel, they faced 20000 Boxers who advanced in a solid mass, their yells deafening and the roar of gongs, drums and horns sounding like thunder. Surrounded, the foreigners could neither escape nor send for help. For almost two months they withstood constant attacks and bombardment. At the eleventh hour, when ammunition, food and medical supplies were almost gone, an international force of soldiers and sailors comprising the eight allies came to the rescue.

The Eight-Power Allied Forces quashed the Boxer Rebellion, looted the capital, and ransacked the Forbidden City, forcing the Empress Dowager to undertake a bold escape from the city as a peasant in a cart. She returned to the Forbidden City a year later, but the power of the Ch'ing dynasty was destroyed forever, and an 'Open Door' policy was forced on China until the Second World War closed it once again.

Haberer arrived in China during the rise of the Boxers and was unfortunate enough to be in Peking at the time of the siege, and subsequent sacking of the city by the Eight-Power Allied Forces. With hundreds of foreigners dead, he must have wondered if he himself would leave China alive. Undaunted, he persisted with his survey of fossils and when he was sent home he took with him a treasure-box full of 'dragon' bones.

Back home in Munich he gave his precious collection to Professor Max Schlosser, a well-known vertebrate palaeontologist. Schlosser worked up Haberer's collection of 'dragon' bones into a monumental publication which (for Europeans) lifted the veil of mysticism over the objects once and for all. It appeared that the 'dragon' bones had no connection whatever with any kind of reptile, but were on the contrary fossil remains of mammals that lived on the Chinese steppes and beside rivers during the past six million years.

Schlosser distinguished no less than 90 mammal forms, divided into widely different groups. The bulk of the collection came from Pliocene deposits (5.2 to 1.64 million years ago) but there was also material from the more recent Pleistocene epoch (1.64 million to 10 000 years ago). The rich Pliocene fauna contained the teeth of bears, different hyenas (one as large as a small cow), otters, sabre-toothed tigers, beavers, mastodons, different rhinoceroses, three-toed horses, hippopotami, pigs, a very large camel, giraffes, different species of deer, and antelopes.

Schlosser's monograph described many species for the first time and was a giant stride forward in the investigation of the ancient life of China. The most controversial specimen in the collection, however, was one of the smallest: a human-like tooth—so much like a human's that Schlosser was able to identify it as the third upper left molar. It was small, very worn and had fused roots. Haberer had collected it from an apothecary shop in Peking. Some red earth was still attached to the roots, and because this condition was common with other Pliocene fossils and the preservation was similar, Schlosser suggested that the human tooth might be of pre-Pleistocene or Tertiary age, that is, older than two million years.

This tooth is wholly fossilised, quite opaque and white yet tinged with reddish colour between the roots, a condition only observed in specimens from the Tertiary. From this fact I am therefore inclined to ascribe to the tooth a Tertiary age. Further the specimen is much worn and also its surface is throughout corroded by the action of plant roots so that no exact opinion can be found as to the original appearance. The general appearance, the contour and the root characteristics of the tooth are man-like, since among anthropoid teeth the roots diverge much more strongly. Only its state of preservation seems to point to a relatively great age, namely Tertiary, for the specimen, and on this account it is somewhat risky to assign this tooth to the genus *Homo*, so long as certain existence of Tertiary man has not been demonstrated. It is necessary therefore to keep in sight the alternative possibility that the specimen belongs to a new anthropoid genus which at least in tooth structure more nearly approaches man than any other known anthropoid.

Schlosser was cautious, and, in his species list, merely described the animal as 'Homo? Anthropoid?' —an indeterminate anthropoid. Some kind of apeman. He hinted that future investigators 'who may perhaps be privileged to carry on excavations in China, that there, either a new fossil anthropoid or Tertiary man, or even ancient Pleistocene man may be found'. This was to be prophetic.

Schlosser's caution was understandable. He would have known that despite the understated description of the find, the tooth would attract much attention in a scientific world still reeling after the discovery in Java by a Dutchman, Eugene Dubois, only eight years before of a possible 'missing link' between apes and Man.

Eugene Dubois was born in Holland to a country apothecary in January 1858. He was a clever student who showed great promise and in 1884, after training in medicine and anatomy, he qualified as a medical doctor. In 1886, he was appointed as lecturer in anatomy at Amsterdam University. A year later, he threw away his career, and enlisted for eight years as Medical Officer Second Class in the Royal Dutch East Indies Army. His aim was to mount a search for human ancestors in the Dutch East Indies, today's Indonesia.

Born the year before Charles Darwin published his *Origin of Species*, Dubois' education occurred during the heady debate surrounding the implications of the evolutionary theory for Man. Dubois' decision to give up a promising, but staid, career was no doubt made under the influence of evolution. Darwin was initially cautious on the question of whether evolution also applied to Man, saying only that 'light will be thrown on the origin of man and his history'. His intent, however, was clear: humans belonged to the animal kingdom and, heaven-forbid, they too evolved, like other animals, by natural selection.

Darwin knew he was undermining the Christian conviction that *Homo sapiens* differed fundamentally from the animals, in the intellectual sense if not in the physical. He remembered the time when people with those ideas had been hanged. It is not surprising, therefore, that Darwin took twelve years to finally come clean, publishing the *Descent of Man* in 1871 and carrying his line of thinking to its logical conclusion that humankind had developed by a gradual process from more ape-like ancestors.

Humans are most closely related to the African apes. Darwin suggested that since apes evolved in Africa, so did humans. Our ancestors were originally tree-dwelling but some shift in the environment, say towards aridity, caused them to come down from the trees to access a broader range of food.

Among the apes we weren't the only ones to climb down from the trees. Baboons, for instance, also became primarily ground-dwellers, but they maintained their quadrupedal form of locomotion, galloping about on all fours. Hominids, however, became terrestrial bipeds. Walkers. Importantly, they became bipeds *before* they became big-brained.

Darwin supposed that the difference between the hominids and other apes reflected the hominid adaptation to hunting. By walking, a ground-dwelling ape would free its hands for carrying tools. As a result, the functions of hominid teeth were partly taken over by tools and so in time our canines reduced in size. Not only did the use of tools lead to a reduction in the size of our canines, Darwin surmised it also signalled the evolution of intelligence.

To many people all this was still an affront. It was one thing to suggest that the Earth was extremely old, and in the course of its history various species had died out leaving fossilised bones, but it was quite another to suggest that human fossils, too, could be found. The Bible taught that Man had only appeared on Earth a few thousand years before, and there was no cause to call biblical authority into question. The fact that no human fossils had been recorded in fossil-bearing deposits up to that time only proved the creationists' point. The very idea of fossil humans was an impossible concept anyway; humankind was of very recent origin and that was that.

Such foot-stamping certainty was, inevitably, undermined throughout the course of the nineteenth century, starting with a series of fascinating archaeological finds. In the 1820s stone tools were found in England in a cave at Kent amid the bones of extinct animals; in 1829 prehistoric habitation sites were discovered on the banks of Swiss lakes and prehistoric middens were found in Denmark; and in the 1840s primitive tools were found in a geological layer yielding up fossils of extinct elephants, rhinoceroses and bears in France.

The idea slowly gained ground that Man was older than earlier investigators had thought possible. So that when Germans Johann Carl Fuhlrott and Hermann Schaaffhausen announced the discovery of Neanderthal Man in 1856, in iceage cave deposits in the Neander Valley near Dusseldorf, there was no problem saying that the bones were ancient.

Fuhlrott and Schaaffhausen even went so far as to say that they were obviously dealing with a hitherto unknown human race (this remains the dominant view today—though debate is fierce). But, having been so bold, their conclusion was unimaginative: they said that the abnormally shaped skull was probably one of the 'savage' races, mentioned by Roman writers, living in the north-west of Europe before the Celts and Teutons. Other anthropologists attributed the bones to a still-extant race, either an old Dutchman, a rickety Cossack, a Celt, or indeed just an idiot. At the time, pathological deformity was often used to explain away strange-shaped skulls: if it didn't look German it was clearly diseased.

Were the bones the first real evidence of human evolution? The question was never asked at that time. The last thing anyone who was interested in evolution studied was palaeontology. As unlikely as it sounds to us today—after all, fossils provide us with the foundations for the sequence of human evolution—leading thinkers of the time thought that comparative anatomy and embryology were all that were required to compile family trees. They simply compared, say, the brain of a Gorilla to that of a human, or observed the development of a human embryo which, they said, would mirror Man's evolutionary stages—from tadpole to fish to mammal. Supplementary evidence from fossil bones was, of course, welcome but it was not actually necessary. The leading German anthropologist and Darwin's disciple, Ernst Haeckel, summed up the attitude of the time:

> We can only partly acknowledge the great importance which laymen and narrow specialists attach to the evidence of such 'fossil men' and 'transitional forms between the apes and man'. He who has thorough knowledge of comparative anatomy and ontogeny, as well as of palaeontology, and who is capable of open-minded comparison of the phenomena, has no need of those fossil documents in order to accept the 'descent of man from the ape' as an historical fact. It appears to us that this is already...a fully founded hypothesis, whether or not later palaeontologic discoveries are going to uncover any 'intermediate forms'.

So unimportant was real fossil evidence of the evolution of Man in the nineteenth century that even Neanderthal Man was an accidental find. No-one mounted a deliberate search for the missing links between ape and Man. No-one, that is, until Eugene Dubois.

By all accounts Dubois was a prickly, precious and contrary character. He was restless and impatient. He was defensive. And he hated teaching. He was, in short, singularly unsuitable for his prestigious position as a lecturer in anatomy, a job that required great patience, precision and dedicated, collaborative laboratory work. One of his colleagues, perhaps too fearful of saying it to his face, wrote this to him:

> You have lost your link with anatomy. From your view of teaching and its value and from your pessimistic ideas of science and from your outspoken dislike of teaching I think that for you, happiness can never lie in theoretical explanation. Posts of this kind are so hard to come by that they ought not to be occupied by people who are not completely contented in this working environment... You knew the work demanded in anatomy and you didn't like it, unless it involved your own research; you liked it so little that you wanted to choose another working environment.

Perhaps it was his contrary nature that forced Dubois to be reactive—deliberately to go against the grain and to go in search of direct proof of the tie between Man and the animal world. To say—as no-one had dared to say—that embryology and comparative anatomy were atavisms. To leave the sophisticated comfort of the Netherlands for the prickly heat of the Dutch East Indies. And to alter our view of ourselves forever.

Charles Darwin had said that the ancestors of Man would be found in Africa because that is where apes, our closest relatives, were found. Apes, said Dubois, were also found in South-East Asia. Fossils of ancient chimpanzees and orang-utans had also been found in the late 1870s, in the Siwalik hills on the border of India and Pakistan, supporting the idea that South-East Asia was the cradle of evolution, not Africa. It was clear to Dubois that the search for the ancestors of Man lay in the Orient. Being a Dutchman, the Dutch East Indies was a logical choice.

It wasn't easy for Dubois at first. Posted to Sumatra, for the first several months he was fully occupied with hospital work and had barely any time to spare for scientific work. After a number of setbacks with the locals who were 'as indolent as frogs in winter' and always trying to put him off the scent, he decided to explore by himself. He did find some productive caves but the effort required him to live in the forest for weeks on end, usually under an overhanging rock or improvised hut, suffering several attacks of malaria. He wrote an article on his finds, which included orang-utans, gibbons, rhinoceroses, tapirs, elephants, deer, cattle and pigs. The article was political. It wasn't so much the animals that were of interest as the fact that fossils could be found in the East Indies in the first place.

Dubois used these finds to appeal to feelings of Dutch national prestige in being at the forefront of research into the natural history of the colonies. He achieved his aim and, in 1889, was seconded to the entourage of the Director of Education, Religion and Industry to carry out palaeontological research in Sumatra and Java. Fifty forced labourers were to help him with his excavation work in Sumatra. Inconveniently, however, some of them died of fever, some ran away and some were sent away for bad behaviour. These were mild irritations to Dubois; by far the biggest blow was the fact that there was barely any difference between the fossil fauna and the fauna of the present day. What was the chance of finding ancient anthropoids that were transitional between ape and Man in modern sediment? Dubois was intent on finding his missing link, so he abandoned Sumatra and transferred the excavations to the Pliocene and Pleistocene-aged sediments of Java in 1890.

In August 1891, Dubois began excavations of a fossil-rich layer of sediment on the banks of the Solo River at Trinil, in eastern Java. His luck was in, and less than a month later, in September, a tooth of a primate emerged from the layer of volcanic sediment. It was a molar—the third molar of the upper right side. Erring on the side of caution for the time being, Dubois designated the find as a 'chimpanzee'. But in October a peculiar skullcap was unearthed. It had a low forehead and jutting eyebrow ridges, like an ape, but a high vault, like a human. From the skullcap it was clear that the Javan chimpanzee was no ordinary ape, being substantially more human-like than any known anthropoid. Dubois knew he had discovered something important. But he needed something else, some other evidence. Frustratingly, the end of year monsoons arrived causing the Solo to flood and work was abandoned as the site was drowned until May 1892. Dubois' prize came in August when an almost completely preserved left thigh bone was uncovered. It lay in the same-aged layer as the molar and skullcap, but 15 metres upstream. Amazingly, it was indistinguishable from a human's.

Dubois was sure that the molar, skullcap and thigh bone belonged to one and the same individual. An individual that was in no way equipped to climb trees. On the contrary it was obvious from the entire construction of the femur that the bone fulfilled the same mechanical role as in the human body. It was clear that the apeman of Java stood upright and moved like a human.

Dubois named his discovery *Pithecanthropus erectus*—the upright apeman—in a provisional account published in January 1894 and written in the East Indies. '*Pithecanthropus erectus* is the transitional form which, according to the theory of evolution, must have existed between Man and the anthropoids. He is Man's ancestor'. These were fighting words!

By 1900, only six years after Dubois had published his description, nearly 80 articles and books had been published in reaction, not including popular articles of which there were countless numbers. Not surprisingly, the reaction was mixed. One German anatomist derided Dubois' work saying he would make short work of his ideas. Of course the femur was a human—no anatomist in the world would look upon the femur as anything but a human thigh bone. But that such a body would have borne an ape skull was a mechanical impossibility. The skull was clearly a new large gibbon species and the thigh bone was that of a human. From across the English Channel, the opinion was that the skull was human rather than ape-like—but it was obviously a microcephalous idiot. A Swiss anthropologist huffed that true scholarship was not well served by the erection of airy speculative constructions. From across the Atlantic, the American view was surprisingly complimentary. They said Dubois had proven to science the existence of a new prehistoric anthropoid, not quite human, but in size, brainpower and erect posture, much nearer man than anything so far discovered.

Dubois retaliated saying that the human-like thigh bone was explicable if one accepted the idea, already put forward by Darwin, that the upright bipedal gait developed before the brain enlarged. As for the disagreement about the skullcap, Dubois countered that the fact that there was such a disagreement of opinion on the skullcap indicated that his interpretation was correct: one expert said it was an ape, another said it was human. Clearly, then, it was an intermediate form and thus a missing link. He had estimated a skull volume of 1000 cubic centimetres which did actually occur in small adult humans. Sure, the fossil femur indicated normal human bodily measurements, and therefore the skull capacity was low—but it was much too high for an ape.

Dubois succeeded in winning a number of experts over to his view. But the critics still insisted that it was not a human ancestor—at best it was a failed attempt at becoming human.

Ultimately, the debate was not about science at all, but about politics and religion. The school that could not believe in any missing link insisted that there was still no proof; only a complete chain of transitional forms between ape and man would satisfy them. For them the argument was against the new dogma of evolution. They implied there was an uncritical following of the theory and that it was not backed by evidence. The paradox was that, against the evidence, they uncritically followed their own dogma of religion and the belief that humans did not evolve and therefore could not evolve. This required some contortions of logic to explain away the human-like thigh bone; humans were clearly created during the early Pleistocene according to their dogma.

By 1900 Dubois had had enough. He had an infinite capacity for annoyance with any scientist whose viewpoint differed from his, and his suspicious nature, verging on the paranoid, caused him to suspect dark motives behind colleagues' disagreements with his views. He become convinced that the Catholics were out to destroy his fossils, so he placed an embargo on them. This had the added bonus of not letting anyone else steal his thunder before his full treatment would come out, which he said he was working on. Perhaps true to form, Dubois never produced a full treatment of the vertebrate fossils of Sumatra and Java.

What was revolutionary was the fact that while half the authors chose not to agree with Dubois' interpretation, half did. They may have disagreed with the detail but they agreed that the *Pithecanthropus* was a transitional form. For the first time in history a body of notable scientists agreed with an evolutionist interpretation. Not only that, but palaeontology became allotted a central place in unravelling the important question of the origin of humankind. Dubois' pioneering work had shifted mountains.

2

Into the teeth of the dragon

The wind swept south across the icesheets of Siberia, chilled further as it climbed the Mongolian Plateau, to whisper and rush through the sparse tundra. It continued south, tumbling through the mixed forests of pines, cedars, elms, hackberries and Chinese redbuds, growing among the uplifting mountains of temperate north-west China, whistling as it funnelled into the ravines and chasms. Toward the southern edge of the mountains the chill wind shivered into a broad valley sprinkled with drying, muddy lakes. The valley lay on the edge of the northern Chinese coastal plain, a rich, vast grass-

land of goosefoot, lily, sagebrush and pyrola stretch-
ing away toward the sea.

The woman shivered as the wind spilled into the
valley. She brushed her unruly hair away from her
low, flat, receding forehead and brow, which jutted
out over her eye sockets like eaves over a house. As
she grimaced toward the position of the sun to see
how much daylight was left, her skin, which
matched the colour of the deep-yellow earth of the
region, crinkled over her high, flat cheekbones and
along the edges of her broad, flattened nose. Her
protruding jaw was chinless, her grimace exposing
her large yellowing teeth. Now in middle-age, she
was already feeling the aches in her bones that
winter would bring.

Her sharp, black, intelligent eyes located some
other members of her clan hiding among the boul-
ders of the valley slopes waiting for the right
moment to ambush the herd. With that wind, it
would be well to restock with skins and meat before
she and her group made their way to more permanent habitation protected from the impending winter
snow-storms. But of course anything could go
wrong. The wind might change direction and the
fire that had been set by others of her clan to start

the horses stampeding might swing in the wrong direction. She glanced about for the children, who would be part of the final chase to mire the horses in the muddy swamp, her eyes scouring the slopes for any big cats, bears, dogs, wolves or hyenas who would easily pick off a small child. Her greatest fear was the sabre-toothed tiger. She had no weapon that could fend off those long, knife-teeth. A lone macaque sat curiously watching the activity of the clan, showing no sign that predators were about. That was good, the woman thought.

Down in the valley she noticed a line of thick-jaw deer making their way to a lake. They had recently migrated into the region, and this autumn migration was another sign that cold was near. From her vantage she could make out the great plains extending to the south-east. In the distance she spotted a woolly rhinoceros, a group of sika deer and a family of elephants. They were looking agitated.

Then she saw a line of smoke rising above the valley wall and heard the fire roar as the wind fanned it. A signal from one of the other clan members told her to be prepared to make the ambush as the horses fled the encircling fire. She focused her entire attention on the opening where the horses would appear and waited for the signal. There it was, a high-pitched whistle. She was aware of the entire clan, springing into action around her, as she sped to circle the horses. She held a club in one hand and a sharply honed piece of bamboo in the other. The children came running behind with rocks. Everyone was yelling and shouting, as if demented. The strongest of the men were pounding directly behind the horses, darting this way and that, on slightly bowed and agile legs, with pointed bamboo-spears urging the horses toward the deadly swamp. The group acted as one, trained by necessity, as the horses plunged into the sticky mud, trapping them like glue.

It had been exhilarating—and bountiful. The clan had managed to trap and kill several horses before the rest of the herd had high kicked and bucked their way to freedom. The woman had sprained an ankle badly but persisted in hauling the dead horses to the butchering site. Pain was part of life, and there was much to do before the scent of the hunt spread to attract the many other predators of mid-Pleistocene northern China. Already the giant hyenas were gathering, the human mothers throwing rocks at them while clutching their small children.

Everyone in the clan had a specific job to do. Some of the men had brought with them a basic tool kit of their best stone scrapers. Mostly made of vein quartz flakes, these scrapers varied in size and shape, their edges straight, convex, concave, or disc-like. Some were multi-edged. Big scrapers and concave edges could be used for shaping hunting clubs and the small ones with similar edges were used to cut into the carcasses of the horses, digging for the nutritious internal organs, and collecting the blood in simple bowls. Other men had already located oval-shaped pebbles of local quartzite in the drying watercourse. A few skilfully executed raps from another stone would see chips fracture and a good double-edge chopper emerge for cutting the wood that the older children were dragging down from the slopes for the fire. An elder of the clan had nurtured a glowing coal for starting the fire. With a well-made chopper it was possible to cut a piece of wood as thick as a man's arm in just a few minutes. The site was already littered with discarded stubs of choppers from previous hunts, which had

been whittled away from use and with further knapping would produce a clean, sharp edge.

Many in the group were savouring the fresh, warm
meat and delicacies of liver and kidney, while others
threw chunks onto the fire, preferring the meat at
least partly fire-seared. The group spoke to each
other about the hunt, some acting out scenes in an
exaggerated manner to the amusement of others.
When the meat was cut into haul-sized pieces the
group collected important belongings, leaving
bones, stones, unusable choppers and scrapers lit-
tering the ground, the hyenas cackling as if in
appreciation of the free dinner they would have from
the scraps. The clan made their way toward a nearby
cave where they would be warm and safe and have
time to process the skins and prepare the meat.

The woman hobbled along behind, finding her
ankle troublesome. The group called to her to hurry,
before darkness fell. She slipped further and further
behind the group, hearing only the loudest shouts
above the strengthening wind which carried the
coldness of the northern glaciers with it. She didn't
hear the soft crack of a twig from above her until
too late, her warning shout to the others cut off as

her neck snapped and the long teeth of the sabre-toothed tiger pierced her chest.

The animal dragged the body back to its cave where it proceeded to eat, breaking and gnawing at the bones, crushing the face and licking at the thick skullcap which was finally discarded at the back of the cave where it lay for eons, buried in the refuse of this and other cave-dwellers.

In 1914, Johan Gunnar Andersson (1874–1960) was prepared to search the length and breadth of China for hints of good deposits of 'dragon' bones. He had carried this passion with him for more than a decade—since he had read the detailed study of fossils from China by Professor Max Schlosser of Munich, based on K.A. Haberer's collection. Officially Andersson was appointed to China as a mining specialist in May, 1914. He was no ordinary mining specialist though and had, ten years before, led a Swedish survey team in the exploration of Antarctica. He was a scientist–adventurer. The fact that no European had ever discovered the sites of the 'dragon' bones was, for him, too alluring for words. Even a tiny fragment of bone would spark his curiosity, and he would spare no efforts to find the bone's source, a hoped-for treasure-trove of unbroken, whole skulls and skeletons of ancient animals.

Unfortunately for palaeontologists, the 'dragon' teeth were regarded by the Chinese as having the greatest medicinal properties, and in order to get to them, scientifically precious skulls and jaws were generally smashed. Whole teeth were often broken as well because the material was very expensive, sold according to weight and used only in small quantities. In this way the pulp cavity was exposed and this often showed small, glittering crystals of calcite due to fossilisation, proof to the lay Chinese that the item was a genuine 'dragon' tooth.

Whole three-toed horses, *Hipparion*, the *index* or pointer fossil of the Pliocene (5.2 to 1.64 mya) were thus smashed up and put in bags, at a time when no museum or research institute in China possessed a complete skeleton. Ancient rhinoceroses were also treated in the same manner when there was not even a lower jawbone, let alone a skull, anywhere in China.

In 1917, Andersson sent a circular to the mission stations in China for help and guidance in his search for 'dragon' bone sites. In the meantime, in early 1918, one of Andersson's friends at the Peking University told him about another fossil site near a town named Chou K'ou Tien only 50 kilometres south-west of Peking, a day's journey on mule-back. Here was a place called 'Chicken Bone Hill', so-named because dense nests of bones of birds were found in its red clay. The possibilities luring him, like sparkling trinkets do a dragon, it wasn't long before Andersson was on a mule on his way to Chou K'ou Tien.

An ordinary village, Chou K'ou Tien was situated on a stretch of land where plain and mountains meet. To its south-east lies the vast North China Plain and to its west and north, the Western Hills. The small Ba'er River emerges from a gorge to the north of the town, winds past the town and joins the Liuli River not far south of the town, which empties into the sea near Tientsin. The seasonally dry Ba'er river bed is over 1 kilometre wide in places and carries rain-fed torrents in summer.

The low hills of the region consist of thick layers of limestone and, on top of that, coal beds. Not surprisingly, the traditional mainstays of the economy for the local people were lime-kiln, coal and other quarry industries. Lime kilns had been established in the region since the Sung dynasty from 960 to 1279. During these times, before explosives were invented, giant iron wedges were used for splitting rocks. The standard procedure was to heat the rock face by means of a wood fire, then pour water over the red-hot face to cause cracks, after which a wedge could be forced into the limestone to break it into transportable chunks. The material was transported by peasants pushing handcarts, or by camels.

When Gunnar Andersson arrived at Chicken Bone Hill he discovered a curious sight—a limestone quarry out of the middle of which rose a detached pillar of bone-bearing clay, 5.5 metres high. It was clear that the clay at one time filled a cavity in the limestone and that the peasants had carefully preserved this mass of clay, which gradually emerged from being a filling in a cavity to become an isolated pillar. A property of the limestone was that any cracks and faults in it allowed water to dissolve away caves and fissures. These were later filled with a red clay and other debris including 'dragon' bones.

Luckily for Andersson, the workers had not touched the bone-bearing deposit. Not because it would have made the lime impure, or because it was difficult to remove (in fact it would have been easy to cart away), but because it was haunted. The story was this.

Once upon a time, more than a hundred years ago, there was a cave here in which lived foxes, which devoured all the chickens of the neighbourhood. In the course of time some of these foxes were transformed into evil spirits. One man tried to kill the foxes, but the evil spirits drove him mad.

Andersson's first 'dragon' bone site, so near to Peking, could not fail to make him happy. The bones, however, were small and he deduced that they most probably belonged to a recent deposit since they resembled still extant animals. So it wasn't a surprise that his interest in Chicken Bone Hill vanished entirely when he received positive responses from his questionnaire to the inland missions of China.

The responses sent him to central China where Swedish missionaries showed him cuttings in some valleys among the cultivated fields. Embedded in the red clay of these cuttings he discovered a 'nest' of fossils 5 metres long by 2 metres high, packed with leg bones, skulls and other bones of ancient animals. Andersson finally held in his hands complete jaws of rhinoceroses and hyenas. It was the first time any European had found a site of real 'dragon' bones.

As exciting as these discoveries were, he knew from the apothecaries in Peking that somewhere else was a site that supplied the medicine market for the whole of China. Through careful sleuthing, he finally found this site in 1919 in a remote and inaccessible place named Pao Te Hsien, far in the north-west of China on the upper reaches of the Yellow River. The early collections from the site soon revealed the fact that here was one of the most important regions so far discovered for mammals of the Pliocene. Not only were there several new types of *Hipparion,* but also many kinds of horned rhinoceroses including the gigantic rhinoceros *Sinotherium lagrelli,* which was entirely new to science. Also new were numerous deer, antelopes, giraffe, elephants and a new group of pigs which were named from a Greek word meaning 'disguise oneself by making ugly grimaces'. Some of the finest material, however, was from beasts of prey including a new genus of cat and three kinds of terrifying sabre-toothed tigers. Andersson quickly realised that he was not up to the task of excavating such a unique site, and entrusted the systematic exploitation of the Pao Te Hsien discovery to what he called 'a real expert', a young Austrian palaeontologist named Otto Zdansky.

Zdansky, a young scholar who had just completed his degree, arrived in Peking in the early summer of 1921 to work for three years under Gunnar Andersson. At a suggestion from Andersson, Zdansky travelled to nearby Chicken Bone Hill in order to get a flavour of working conditions in the more remote Chinese country districts he would be working in later. After making a base in a dilapidated village temple at Chou K'ou Tien, Zdansky, together with Andersson, sat down to scratch the fossils out of the haunted pillar of clay. Not long after they began their work, a local man came to look at what they were doing, after a while commenting that they were wasting their time because just a few hundred metres away from Chicken Bone Hill was a place

Initial digging at the Peking Man site, 1921.
Otto Zdansky is on the left.

where they could collect much larger and better 'dragon' bones.

Since it seemed clear the man knew what he was talking about, Zdansky and Andersson packed up their kit and followed him over limestone hills to an abandoned quarry on what was called Dragon Bone Hill, 150 metres west of the Chou K'ou Tien railway station. In an almost perpendicular wall of limestone, about 10 metres high, the local man revealed a filled-up fissure in the limestone consisting of pieces of limestone and fragments of bones of larger animals, cemented together in a red clay matrix. It seemed that, originally, there had been a complicated cave system, now filled with sediment and bones. After only a few minutes Zdansky retrieved a jaw of a pig, and, the following day, rhinoceros teeth and jaws of hyena and bear.

The sediment and bone had been deposited under calm conditions and were stratified into distinct layers. Andersson noticed that two of the eight or so layers contained angular pieces of quartz with sharp edges, almost as if they had been worked by forces other than wind and water. Indeed, the edges were sharp enough to cut meat. Tapping on the wall of the limestone, he mused to Zdansky, 'I have a feeling that there lie here the remains of one of our ancestors and it is only a question of your finding him. Take your time and stick to it till the cave is emptied, if need be'.

In the event, this proved to be an almost impossible task for Zdansky without large scaffolding to stop the slabs of sediment falling on top of the excavators who were beavering underneath the now overhanging wall of the cave deposits. Zdansky eventually called it a day, concluding his investigations in the late summer of 1921 and turning his attention to the rich *Hipparion* fossil layers of central China.

Andersson, however, could not shake the feeling that among the badly broken bones of sabre-toothed tiger, bear, horse, dog and deer at Dragon Bone Hill lay the remains of man. At Andersson's request Zdansky returned to Chou K'ou Tien in the summer of 1923 and began once again to carefully excavate the broken bones of Dragon Bone Hill. It was becoming increasingly dangerous without scaffolding and it wasn't long before he was forced, a second time, to give up his search due to lack of funds to pay for scaffolding. Nevertheless, Zdansky still managed to extract a variety of animal bones including a peculiar molar belonging, he thought, to an ape. Not much significance was attributed to the tooth and it went to Andersson's collaborator Professor Wiman of the Museum of Uppsala in Sweden together with a great quantity of fossils—enough material for years of laboratory work.

In October 1926, the Crown Prince and Princess of Sweden, who were doing a round-the-world tour, asked if Andersson could organise to inform the Prince, while he was visiting Peking, of recent discoveries in China. The Prince, an amateur archaeologist, was also the Protector of the Swedish Research Committee who was financing Andersson's adventures in China. Andersson, wanting to be on the right side of his benefactor, decided to organise a special scientific conference for the Prince, where Peking scholars could elaborate on their latest work. For his own research, Andersson wrote to Professor Wiman and asked him to send the latest descriptions of the Chinese mammal material. Wiman wrote back immediately with the latest discoveries giving an account of a new dinosaur which, he said, 'was perhaps our most valuable discovery', and of the peculiar giraffes of the *Hipparion* deposits, as well as of the snouted horse, *Proboscidipparion*.

Almost in passing, he mentioned that Zdansky had recovered a premolar resembling that of a human being, while cleaning the Chou K'ou Tien material at Uppsala. The premolar had made Zdansky reconsider the so-called 'anthropoid ape' (gorilla, orang-utan, chimpanzee or gibbon) tooth he had recovered at Chou K'ou Tien in 1923, so that he now designated both teeth as *'homo* sp?', the question mark indicating that he was not prepared to stick his neck out.

Andersson was. He thought that the incomplete discovery was likely to be the most important result of all the Swedish work done in China. Saving the best for last, as the final speaker at the Prince's scientific conference, Andersson announced the astounding discovery of two molars of a creature resembling a human being.

The teeth were turned over to Dr Davidson Black, head of the anatomy department of the Peking Union Medical College and an expert on dental evolution and comparative osteology (the study of skeletons), for examination and further investigation.

In the history of the study of human evolution there is a series of associations that have become fixed: *Pithecanthropus erectus* from Java and his discoverer Eugene Dubois; Raymond Dart and *Australopithecus africanus* from southern Africa; Louis Leakey with east African *Homo habilis*. If there is one name associated with the discovery of Peking Man, *Sinanthropus pekinensis*, it is Davidson Black.

Davidson Black was born on 25 July 1884, in Toronto, the son of a lawyer. As a boy he spent his summers pursuing his love of exploration and adventure on the Kawartha Lakes of Canada. His skill as a canoeist and love of nature enabled him, during high school summer vacations, to take a job hauling supplies by canoe for the Hudson Bay Company. These solitary trips through wild and uninhabited country tested the self-reliance and skill of the young Black. One anecdote has it that he escaped an encircling forest fire by standing, immersed to his neck, in a lake for two nights and a day.

He maintained his summer work in the back woods of Canada, nurturing his love of nature and interest in geology, while pursuing a medical degree at the University of Toronto. In 1906, at the age of 22, Black became a doctor and, with the letters M.D.

Davidson Black, the first to
recognise the identity of
Peking Man.

after his name, promptly decided that medicine as
a career did not appeal to him after all. He returned
to the University to study his first love, biology,
reversing the order in which arts and medicine is
usually taken and receiving a Bachelor of Arts
(in Biology) in 1909. That same year he accepted
a lectureship in the department of anatomy at
Western Reserve University in Cleveland, Ohio. He
still spent his summer holidays doing fieldwork in

outback Canada, by now a competent amateur geologist, a topic he studied in his spare time. The logic
of Black's career was not easily perceived by the
casual observer. A friend wrote 'one was always
coming across new and unexpected topics to which
Black proved to have given attention at some period
in the past'.

His most exacting studies, however, were in human
anatomy and in neuroanatomy. In 1914, Davidson
Black travelled abroad to study neuroanatomy with
the famous Australian, Sir Grafton Elliot Smith, who
was working in England. At the time, Smith was
working on the reconstruction of the newly found
Piltdown Man, another contender for the title of
'missing link'. Smith was studying the pattern of the
brain on the inside of the skull and, for the pur-
poses of comparison, had collected casts of all the
known fossil human skulls. This work aroused a
much greater interest in Davidson Black than the
brains of the lungfish that Smith was trying to
persuade him to study.

Black was captivated with the problems of human
evolution that the Piltdown skull represented. It is
ironic, to say the least, that the Piltdown skull,
which catalysed the sea-change for Black, turned out

to be one of the most famous frauds in the history of science. It is perhaps just as well that the scientifically rigorous Black never knew it.

The main actor in the Piltdown Man drama, until his death in 1916, was Charles Dawson, a solicitor and an amateur archaeologist and collector of fossils for the British Museum. From 1908, Dawson had been collecting the fragments of fossilised bone which were being thrown up by roadmakers near Piltdown in southern England. By 1911, Dawson had collected enough pieces to convince himself that these were indeed remains of a primitive human being. In 1912, he took his collection of fossils to Arthur Smith Woodward, the Keeper of the Department of Geology in the South Kensington Museum, who declared that they belonged to an ancient period. Later the two men examined the pits together and found other pieces of a skull that seemed to be part of the original find, and later a monkey-like jawbone with a few teeth. The excitement caused by these finds was great and the newspapers took up the cry that Dawn Man had been found in the south of England. Scientists, however, disagreed among themselves as to how a monkey jawbone could be made to fit with an obviously human skull. No-one suspected that the finds were anything but genuine.

At the time of the Piltdown finds there were very few early hominid fossils. Neanderthal Man was considered to be human, and there was continuing heated debate about the 'transitional form' status of Dubois' *Pithecanthropus erectus*. It was expected that there was a 'missing link' between ape and man but it was an open question as to what that missing link would look like. Piltdown Man had the theoretically expected mix of features—a mixture of human and ape with the noble brow of *Homo sapiens* and an ape jaw—which lent it credibility. As did the fact that it was English!

During the next two decades, however, Piltdown Man became a problem child. It did not fit in with new discoveries. As a result, it was increasingly marginalised, and eventually, simply ignored; one palaeontologist wrote that 'you could make sense of human evolution if you didn't try to put Piltdown Man into it'. Piltdown Man was, however, carried in the books as a fossil hominid, puzzled over from time to time, until being dismissed once again. Then in July 1953, an international congress of palaeontologists examined the world's cache of fossil hominids. Against such a line-up, Piltdown Man looked suspicious. He simply did not fit in with the crowd. He was a crooked piece in the puzzle. It finally dawned on one of the scientists that perhaps he was a fraud.

Once the possibility had been raised, so were the mists of uncertainty surrounding the Piltdown fossils and it became ridiculously easy to see that the finds were a fake. For a start, another look at the teeth showed that they had been filed to fit: the first and second molars were worn to the same degree, and the inner margins of the lower teeth were more worn than the outer. In other words, the wear was the wrong way around! Inspection under a microscope immediately confirmed the artificial abrasions.

How was it that, for 40 years, the fraud had escaped notice? The answer was simply that nobody had previously examined the Piltdown specimens with the idea of a possible forgery in mind.

If the answer was simple, the forgery itself was simply breathtaking. In the first instance, there were never any significant fossils at the Piltdown quarry. Instead it was salted from time to time with bones from a variety of sources: a medieval human skull;

a orang-utan jaw from Sarawak; a Pleistocene chimpanzee fossil canine; a fossil elephant molar from Tunisia; and a fossil hippopotamus tooth, from Malta or Sicily. The latter two teeth were meant to indicate the fauna during Piltdown Man's era.

Not only were the bones gathered from a variety of sources, they were given a thorough treatment to make them appear to be genuinely ancient and emulate natural conditions. Fossil bones naturally deposited in gravel pick up iron and manganese. Therefore a solution containing iron was used to stain the modern bones, after they had been treated with chromic acid to facilitate the iron uptake. The canine tooth was painted brown after staining, and the jawbone molars were filed to fit. The connection where the jawbone would meet the rest of the skull was broken off so that there would be no evidence of lack of fit. The canine tooth was also filed to show wear and its hollow centre was filled with sand as it might have been if it had lain in a river bed for eons.

Amazingly, when the hoax was exposed nobody had a clue who the perpetrator was. No-one confessed. And many of those involved were now dead. For 40 more years, right up to the present day, people have speculated about the identity of the culprit. Over time an impressive list of suspects has accumulated. The case against each has been circumstantial, a fantasy of suspicious behaviour, of possible motives, and of opportunity; a who's who of whodunits.

Perhaps not surprisingly, one of the suspects was Grafton Elliot Smith. It was said he had the right kind of personality (he was an Australian after all). It was Smith who brought Davidson Black down to the Piltdown quarry in 1914. Interestingly enough, Black noted that, based on the geology of the place, ancient man could not have lived there, suggesting the bones may have been washed in. In an article titled 'Brain in Primitive Man' written not long afterwards, Black stated: 'All evidence of man's antiquity and of his primitive structure and civilisation depends upon the natural preservation in geological strata of known age, of certain hard parts of the skeleton and teeth and of certain stone implements of human structure'. While not writing specifically about Piltdown Man, it is a pity no-one took Davidson Black's insight to heart.

The First World War was declared while Black was in Britain and he returned to Canada where he immediately offered his services to the Army who rejected him because of a slight heart murmur. The noble Black felt this keenly. And the unexciting humdrum—Black called it 'the daily round, the common task'—of Western Reserve University life could not replace the intellectual excitement of Europe. Black escaped by burying himself in books. In that frame of mind it was no surprise that a book called *Climate and Evolution* by William Matthew, which appeared in 1915, would capture Black's full attention.

Matthew's thesis was that Asia, not Europe or England, was of paramount importance in the evolution and dispersal of primates. In ideas that encapsulate the thinking of the time Matthew wrote 'most authorities are today agreed in placing the centre of dispersal of the human race in Asia', naming the great plateau of central Asia as the likely area. Matthew pointed out that immediately around the borders of the great plateau are regions of the earliest recorded civilisations, in Egypt, India and China. Matthew supposed that waves of invasions from central Asia pushed civilisation to the corners of the earth along mountain paths and down lush valleys. From this perspective Australasia, including

Australia, south-western era and western Africa, and northern Brazil were most remote in terms of travel-routes and therefore these lands contained 'the lowest and most primitive races of men'.

Matthew did not think much of the idea that man was primarily adapted to a tropical climate. After all 'it will not be questioned that the higher races of man [that is, in Europe] are adapted to a cool-temperate climate, and to an environment rather of open grassy plains than of dense moist forests. In such cool climates they reach their highest physical, mental and social attainments'. Even in Africa, Matthew pointed out, the negro reaches his highest physical development not in the jungles but in the drier and cooler highlands of east Africa, 'and when transported to the temperate United States, the West Coast negro yet finds the environment a more favourable one than that to which his ancestors have been endeavouring for thousands of years to accustom themselves'.

Black must have taken much of this with a grain of salt, but the idea that central Asia was the centre of evolution and dispersal of man was, to him, inspiration itself. The spark that had been lit with Piltdown Man now burst into flame. Black wanted to go to China to explore the origins of humanity. One of his closest friends observed the effect:

Under the spell of the tremendous fascination of that problem was born the greatest of the several personages of Davidson Black, namely Davidson Black the Anthropologist. Here was the call from the unknown for which his restless spirit had been waiting; a call that came as a challenge and an opportunity; one of those rare opportunities that come only to the prepared.

There can be no doubt that from this time on the problem of the origin and early evolution of man occupied first place in Black's mind. Immediately the scattered rays of previous interests came to a clear focus on this fascinating subject. He recognised at once that here was an opportunity for full investment of his unique capital in life: his inherent love of adventure, his instinct for discovery, even more his practical geological experience and perspective and his extensive knowledge of comparative anatomy in particular. All his previous interests and experiences immediately fell into their self-appointed places in this broad foundation of correlated qualifications that guided his approach to the new problem now uppermost in his mind.

In 1917, the United States declared war on Germany and Davidson Black once more presented himself for military service. This time, at his earnest insistence, he was accepted and given the rank of captain in the Army Medical Corps where he worked in England in Canadian hospitals that were taxed to the limit by the tremendous stream of casualties from the western front.

As fate would have it, on the eve of his departure for England, Black received an excited letter from a friend of his student days, Dr Vincent Cowdry, who had just accepted an appointment in China, as head of the Anatomy Department at the newly established Peking Union Medical College, a philanthropic enterprise of the Rockefeller Foundation. Cowdry asked Black if he would accept a position in Peking, at the Medical College, as Professor of Neurology and Embryology, working under Cowdry.

Here, finally, was the chance Black had been waiting for, in the country on which he had set his heart—on the eve of his departure to fight in the First World War. His frustration was relatively short-lived, however.

Black arrived in Peking in August 1919 to take up his work as Professor of Neurology and Embryology at the Peking Union Medical College. The Rockefeller Foundation, thankfully, had accepted his appointment on the understanding that he would take up his duties after completion of his military work.

Initially Black's activities within the College focused on obtaining 'material' for his research.

I am having great fun in getting the material here. This week I bought eight monkeys, all Macaques. Three from southern China and five from Siam [Thailand]. During the last month I have had a large female camel living in the basement. Read, our physiological chemist, is doing the nitrogen metabolism on this beast and I shall take what remains. The experiment will probably continue from six to eight weeks. It is a treat to see this camel objecting to coming indoors. The whole neighbourhood is made aware of her objections. I may remark in passing that I have come across some Mongolian camel breeders and hope to be able to get some embryos and foetuses as well as obtaining some information with regard to their breeding habits.

My comparative osteological material is accumulating fairly rapidly... The trouble with collecting Primate specimens is that just at present I have to buy these animals from dealers who sell them as pets at a much advanced price. I paid $8.50 for each of my Macaques and they are all very small individuals. When I go into the field next summer I hope to be able to collect some Primate stuff at a more reasonable figure.

I am going to East Manchuria next summer to investigate a number of cave burials of which I have heard from a Jesuit missionary there. I have great hope of getting some good stuff but I shall be glad to get any material. It is much more difficult to obtain human material for study than it is to obtain subjects for execution! The outlying authorities do not seem to have any special regard for human life but they are not at all enthusiastic about having material used after death; however, we have hope.

While the letter was light-hearted, it hints at the anarchic circumstances in China during its long civil war. Perhaps it was only in such a light-hearted manner that life could be lived in Peking. Another anecdote Black told his colleagues followed a question, while visiting Canada, about whether he found it hard to get cadavers for dissection. Black admitted the difficulties since Chinese of all ranks worshipped the dead and were unwilling to have them used in this way. To overcome this, it was suggested that Black apply to the local prison for help. In due course three bodies arrived but they were headless. It was tactfully pointed out that they were of no use in this state, so when the next lot came they arrived on their own feet bearing a note inviting Black to 'kill them any way you like'!

It was against this bloody background of 'disregard for human life' that the equally bizarre fantasy life continued among the foreigners who continued to pour into China's great cities. The numerous legations in Peking reacted to the civil war simply by augmenting their staffs to deal with the problems of their nationals, and to help foster any trade that might be advantageous to their countries. Also, the western-educated and scientifically trained Chinese welcomed the opportunity to associate with fellow scientists on the staff of such institutions as the Peking Union Medical College. This coming and going of men with scholarly interests and varied nationalities made life stimulating and the genial, intellectual atmosphere, where cocktail and dinner parties were the order of the day, was in sharp contrast to the constant warring of the Chinese political parties.

In the early years of the twentieth century, the ancient imperial, Confucian order finally collapsed with the demise of the Ch'ing dynasty after the Boxer Rebellion. Pressure from the West, China's struggle for modernisation and her ongoing internal warring activities, all came to a head. Unfortunately there was nothing to replace the old order—which had been in existence since the Western Han period (206 BC to 9 AD)—so the Chinese military leaders, inevitably, turned to force to resolve their disputes.

From 1911 to 1949, China was run by warlords who were adept at devising new taxes, at collecting land taxes often many years in advance, at promoting the cultivation and sale of opium and in extorting protection money. Nepotism reigned, and any general who became a warlord would promote his relatives and subordinates who had aided him in his rise to power.

At the bottom of the power pyramid soldiers were recruited by poverty. Even bandits joined the warlord armies, as individuals or even as whole gangs. Soldiers' pay was at or below subsistence, and medical services ranged from poor to non-existent, so that even a superficial wound in the often bloody battles between warlords had a good chance of resulting in a painful death.

Foreigners, such as Davidson Black, who looked at this period of high warlordism from the privileged and relatively safe vantage of the treaty ports often viewed the warlord wars as less than serious. The battles were usually preceded by endless political manoeuvring and outrageously mendacious public statements, and were usually decided by betrayal. Armies of soldiers would break off the fighting because of rain, lunch or nightfall.

As well as being corrupt and corruptible, warlords were also eccentric. One warlord, Feng Yu-hsiang, in his Christian phase, baptised his entire brigade using a fire hose to ejaculate the holy water. In a subsequent communist phase, he would abandon his normal ornately gold braided and bemedalled marshal's uniform, and clothe himself in the quilted cotton worn by common soldiers. Another warlord, Chang Tsung-ch'ang, was always accompanied by his harem of White Russian concubines, but earned the name 'dog meat general' from his culinary appetites. He was also nicknamed 'three don't knows' since he didn't know how much money, how many troops or how many women he had. In utter contrast, a classically trained warlord, Wu P'ei-fu, could be seen on the battlefield in a scholar's gown, with brush and inkstone, composing poetry. Not surprisingly he was the favourite of the British, who also liked the warlord of the nearby Shansi province, Yen Hsi-shan, because he made the trains go on time.

Against this background the Peking Union Medical College continued to hold its dinner parties and evening soirées. In an oasis protected by the ample funds of the Rockefeller Foundation, the scientists pursued their teaching and studies, unperturbed.

But between 1919 and 1925 the warlords clotted into competing military allegiances. Wars between these factions were on a much larger and more ferocious scale. At the same time, other centripetal forces were at work within China. Sun Yat-sen, the charismatic socialist, was working toward an American-style republic and, from his base in Canton, in the south of China, he was gathering together his National Revolutionary Army to subdue the warlords of the north. His hopes to join the north and south of China were not realised when he died of cancer at the Peking Union Medical

College in March 1925. His brother-in-law Chiang Kai-shek, took his place. A soldier by training, rather than the diplomat that Sun Yat-sen could be, Chiang Kai-shek amassed a huge army and marched north to overthrow the powerful northern forces. His real ambition, however, was to subdue any opposition and make himself dictator of China.

In 1925 Davidson Black, himself now head of the anatomy department after the departure of Vincent Cowdry, wrote a letter to his new assistant professor, Dr Drooglever Fortuyn, while he was en route to China to take up the position. Black explained that 'although the country is always in a politically unsettled condition we rarely suffer any inconveniences and conditions for work are almost ideal. During the next teaching year I hope you will undertake the responsibility of giving the course in histology and microscopy'.

In fact, unimaginable anxieties lay ahead for the Fortuyn family, who nevertheless stayed on until 1942 when, as internees of the Japanese, during the Second World War, they were released on an exchange basis. Fortuyn was more sensitive to the situation in China than Black. He wrote in his diary shortly after arriving in Peking:

October 24, 1925: The political and military situation looks to us like a game of chess. Everyone moves his troops and whoever has the strongest position wins, without any real fighting.

But then the tone changes:

December 21, 1925: A serious battle seems to be in preparation and so we gave money for the care of the wounded. It is strange how little such things affect ourselves and other foreigners.

December 24, 1925: Last night there was a Christmas party at the Peking Union Medical College. It began with a not exactly peaceful story of Professor H.J. Howard being kidnapped by bandits last summer.

March 11, 1926: It is nearly impossible to do anything for the Chinese, because there is no security. Politically speaking we are not in a revolution but in a state of anarchy. There is practically no government and no ruling class.

March 21, 1926: We learned this afternoon that the ministers in Peking of the countries with extraterrestrial rights may expel their citizens from China without explanation. Therefore if tomorrow our Netherlands minister says we must leave China—we must. A peculiar but not incomprehensible situation.

March 27, 1926: The situation is again critical. Our legation urges us to go from our dwelling compound to the Peking Union Medical College at the hoisting of the Blue Peter on the radio mast of the U.S. Legation and from the College soldiers will conduct us to the Legation quarters. This means everything must be left except a few necessities, but tomorrow we shall hide our silver in the College.

These diary entries coincided with events in the cities south of Peking. Here the National Revolutionary Army, which was now an uneasy alliance between Chiang Kai-shek's National Party, and the mushrooming Communist Party, was occupying Hankow, Shanghai and Nanking. The occupation had been followed by communist-led mass action directed against foreigners. This action had forced the British, under duress, to return their authority over their concession areas to the Chinese.

Sun Yat-sen, before his death, had not been oblivious to the growing Communist power in his party but he was willing to take the risk because the Communists brought with them money and support from Russia. Chiang Kai-shek was not so patient. On the nights of 20 and 21 March he raided all Soviet quarters in Canton and confiscated all arms, essentially stripping the Communists of power. It was perhaps fear of a war against foreign superpowers such as Britain and the Soviets, that sent nervous ripples throughout other foreign offices and put all foreigners on red-alert.

Meanwhile, unrelated battles between the warlords of northern China (who were funded by an acquisitive Japan) kept the community of the Peking Union Medical College on their toes, as reflected in Fortuyn's continuing diary entries:

April 3, 1926: Bombs dropped on Peking. You will be very much alarmed by this and I don't understand myself why we remain so calm but everything goes on as usual and so you act yourself as if nothing happens.

April 17, 1926: Politics are again quiet. A 'Safety Committee' will try to keep the troops outside the city. The gates of Peking are now really closed and foreigners have had to climb over the city wall by means of a rope.

April 19, 1926: After dining in the Hotel de Peking we went home about eleven o'clock and heard gunfire around Peking which continued all night. It stopped in the morning because, we were told, the Chinese prefer to shoot at night—you have less chance of hurting anyone and you can see so beautifully the flame leaving the guns!

After these alarms the fighting moved away from Peking, although there were still troop encounters near Chou K'ou Tien.

It is difficult to picture serious academic life and archaeological research being carried on under such conditions. But, until war became worldwide, it did.

The last thing on Davidson Black's mind was the squabbling of warlords when Gunnar Andersson announced the discovery of the possible hominid teeth at the Prince of Sweden's scientific conference in Peking in October 1926. With the teeth in his possession Black penned an excited note to a colleague:

> There is great news to tell you—actual fossil remains of a man-like being have at last been found in eastern Asia, in fact quite close to Peking. This discovery fits in exactly with the hypothesis as to the Central Asiatic origin of the *Hominidae* which I reviewed in my paper 'Asia and the Dispersal of Primates'. I have every hope that I shall shortly be able to organise a systematic two-year research project on the Chou K'ou Tien deposit.

After examining the teeth, Black was so confident that they were hominid that he even identified them as relics of a new type of early man going back to the early Pleistocene, each possibly to a million years or more.

> For the first time on the Asiatic continent north of the Himalayas archaic hominid fossil material had been recovered accompanied by complete and certain geological data. The actual presence of early man in eastern Asia is therefore now no longer a matter of conjecture...
>
> The Chou K'ou Tien molar tooth, though unworn,

would seem to resemble in its essential features the specimen purchased by Haberer in a Peking native drug shop and subsequently described in 1903 by Schlosser. It was provisionally designated *Homo? Anthropoid?*. It is of more than passing interest to recall that Schlosser, in concluding his description of the tooth, pointed out that the future investigators might expect to find in China a new fossil anthropoid, Tertiary man or ancient Pleistocene man. The Chou K'ou Tien discovery thus constituted a striking confirmation of that prediction... furnishing one more link in the already strong chain of evidence supporting the hypothesis of the Central Asiatic origin of the *Hominidae*.

Peking Man was born.

3

Peking man

Two million years ago—the beginning of the Pleistocene Epoch—was a time to remember. It was a time when two of the world's largest rivers (the Yangtze and the Yellow River) came into existence, streaming off the recently risen Tibetan Plateau, now the roof of the world. Ironically, for all the antiquity of its culture, the main features that give character to the very face of China itself, like the Tibetan Plateau, are relatively young. A gargantuan feature—easily the size of western Europe—the Tibetan Plateau came into existence with the collision of India into China. It was a profound reorganisation, gradual for the first 20 million years; then, on an earthly scale of things, virtually tumbling skyward in the last 2 million years to altitudes of over 5000 m.

Global in its impact, the tremendous increase in the elevation of such a slab of the planet changed even the swirling pattern of the atmosphere. In China, a benign, moist monsoonal climate, which nurtured herds of migrating animals, switched to that of a desert. The upward thrust of the Tibetan Plateau not only became a barrier, blocking the summer monsoonal air sweeping into India and causing desert conditions in north-western China, it also pushed the Eurasian High Pressure System, which originally squatted over China, north over Siberia, from where the strengthening winds howled back into China. The result of the Tibetan uplift tipped the Earth over the edge into an iceage.

Nothing could be further from the truth than to consider today's climate normal. The last time conditions were like they are today was 125 000 years ago, and then only for a short period lasting maybe 4000 years, during an interglacial.

The current (Pleistocene) iceage is characterised by regular glacials punctuated by interglacials (one of which we are currently in). During the glacials, rivers dry up, temperatures plummet and with them the capacity of the atmosphere to hold water. Under these circumstances glacials coincide with periods of aridity; the warmer interglacials with higher sea levels and more rainfall. Relatively warm and wet peaks rapidly plummet to extremely cold and dry troughs, and back again. These extremes are restricted to brief peaks, but the overall cooler temperatures extend over a greater length of time than before the Pleistocene, and so produce an overall decline in rainfall.

In China, the strain of these conditions are etched on its face, giving much of the landscape its character and its name. Huang T'u, the yellow earth, is the name given by the Chinese to the dust-fine soil which rules north China during arid times. Western science calls it 'loess' because a similar soil in the Rhineland has been thus named.

Loess grains are mainly transported from the inland deserts in north-western China by the winter winds of the Siberian High Pressure System. When extensive sheets of ice exist, the high pressure system is strongest and the wind fiercest, stripping the sand and silt from bare earth, lifting it up to 10 kilometres and transporting it in the jet stream for distances of 1500 kilometres or more before blanketing the landscape where the velocity of the wind is least and the curves and hollows of the ground offer shelter and protection. Even today, several centimetres of silt may be dumped in a single violent dust storm.

Where the yellow earth prevails, it rules over the air, land and the waters. It fills up the valleys, above which the mountain ridges rise in much the same way as mountains emerge from winter snow. A thing of cruel beauty, an unbroken cover of loess disintegrates into a bewildering topography of narrow ravines with perpendicular walls, detached islands, castle-like pillars and frail, skull-like vaults and arches.

In China, the alternation of loess and moister soils mirror, with astounding accuracy, the oscillations of the iceage between a bare or poorly vegetated semi-desert on the one hand, and grasslands, woody grasslands, and woodlands on the other. This geological layer-cake begins at a time coinciding with the first glaciers in Europe, just before 2 million years ago.

At the same time in Africa, a revolution was bubbling away; a revolution that would spill over to Eurasia and, more destructive than the world-engulfing iceage, would eventually change the natural order of things forever. It was the human revolution. Two million years ago, the main species of hominid—the family comprising man and his man-like ancestors—was tall and rangy with long legs. It had lost its fur and had evolved improved water retention and the ability to sweat. It was a communal animal—hunter-gatherers that followed the population expansions and dispersals of herds of grazing ungulates.

Humans had a complicated social system where sexual division of labour allowed family groups to use different resources of any one region, cleverly increasing the carrying capacity of the land. This had the spin-off of allowing groups to expand into new and more difficult habitats. Groups of animals adapting to hunting, scavenging and gathering require larger home ranges in more difficult habitats to take best advantage of seasonal resources. This meant that groups would necessarily come into regular and predictable contact—in places where rich resources of seasonal fruits were ripening for instance

—hence developing broad social networks. This would, in turn, broaden the knowledge base of new habitats, allowing places to be exploited with fewer risks.

In addition, humans had a tool kit. While tools had been used by hominids for millions of years, here, for the first time, the tools were preconceived by their makers. More than just banging rocks together, here was an animal that would gaze at a rock and conceptualise the symmetry and form of tool that could be carved from it. The symbol of this conceptualisation was the hand-axe, a pear-shaped tool that was bifacially flaked to form an edge, and pointed at the tip for general purpose use. The pen-knife of the palaeolithic, hand-axes were of vastly differing sizes and used for cutting, digging, chopping, wedging and drilling. More than a simple tool, the evolution of the hand axe, which continues to be used right up to the present time, mirrored the evolution of a human mind.

Numerous new types of tools also appeared. Flakes were reworked and occasionally sharpened and changed from one tool type to another. The retouch process revealed human insight about the characteristics that tools needed to work effectively. People began to take care and curate the tools that they made. Paralleling the development of tools was a shift in human ranging and foraging behaviour. Higher, drier areas with open scrub and grasslands were colonised. People began to move away from permanent river and lake camp sites to seasonally ephemeral streams. It was just as well, because it was these drier habitats, where the greatest concentration of game was found, that would characterise much of the human habitat once the iceage was ushered in.

During glacial periods, the Sahara shifted from a wet and fertile place to a desert. It may have acted like a pump drawing in populations when it was luxurious and spitting them out (presumably right out of Africa) when it was arid.

At that time, however, humans were still part of the overall natural community. Their first migration out of Africa was probably part of a more general event since the hippopotamus, the forest elephant, the lion, the leopard and the spotted hyena also migrated northwards from Africa at the beginning of the Pleistocene. The fact that conditions must have been advantageous for large predators at this time is indicated by the size of migrating carnivorous species. The first lions, leopards and spotted hyenas that made Europe their home were enormous, a feature suggesting good times and an abundance of large game. These large predators replaced the more inefficient Eurasian carnivores and scavengers, opening a niche for hominid scavengers.

This, of course was not the first dispersal of species across continents. About 19 million years ago, Africa, which until then had not been connected to Eurasia, finally drifted closer. Like a boat clumsily pulling alongside a jetty the collision was really a series of bumps. The first bump was between the Arabian and Turkish regions of Africa and Eurasia respectively. While the continents grappled with one another, ancient carnivores, pigs, bovids and rodents scuttled across from Asia to Africa, while primates, elephants and *creodonts* (a group of ancient land-dwelling carnivores which gave rise to the whales and dolphins) passed from Africa to Asia. The continents then parted, only to briefly dock again around 16 million years ago when newer groups of primates, pigs and elephants boarded Eurasia. The final docking—12 million years ago— saw the rhinoceros, hyena and sabre-toothed cats disperse into Eurasia.

About 2 million years ago, humans became a colonising species, expanding their ecological range into arid and highland-to-mountainous habitats, and eventually moving out of Africa to spread across the tropical and subtropical regions of the Old World, where repeated lowering of the sea level by 200 to 300 metres during the Pleistocene glacials allowed herds of animals and groups of humans to walk into Java and Borneo across land.

Human fossils, dated at 1.8 million years, have been found in Israel and Georgia—exactly where one would expect humans to be on the exodus from Africa. They are associated with other fauna that represent migrations. Human populations had begun to inhabit other regions in numbers significant enough to leave an archaeological and fossil record. Indonesian woodlands, too, were home to some of the earliest humans found outside of Africa. Human remains possibly as old as 1.8 to 1.6 million years, are the earliest found here, indicating that colonisation of the Old World occurred quite rapidly. The earliest convincing use of fire is at 1.6 million years ago. Populations inhabited the tropics of South and South-East Asia, and ranged eastward into central China and as far west as the western edge of Asia.

At this time, the Tibetan Plateau was still a forested expanse, and no real barrier to migration, and much of China was subtropical forests.

The critical changes that accelerated the colonising process were social ones, as broad networks reflecting interactions and communication between populations provided the chance for populations to take advantage of new opportunities with fewer risks. The rapid spread of the tool kit attests to the networks, just as the expanding human range reflects the consequences of the networks.

While the human revolution was occurring in Africa two million years ago and the Tibetan Plateau was poised to convulse skyward, northern China lay under a moist, subtropical sun. The warm summer rains pattered on a gently undulating plain, in the region around where Dragon Bone Hill now protrudes, seeping and sliding along the steeply tilted layers of limestone that underlie the plain. Resistant to anything but water, the rock gradually dissolved into the fluid embrace, forming a hollow and then an underground cave.

Monstrous stresses jostling at the very foundations of the rock caused the plain to buckle gently into a hill. A river flowing against the eastern slope of the rising hill ate at one of the walls and eventually bit away a small entrance to the cave. Once breached, the water bubbled into the cave, depositing pebbles and layers of sand over the rugged, limestone-sharp ridges, smoothing out the cave floor. Still not comfortable in its position, the hill continued to rise, lifting the enlarging entrance of the cave above the daily river-flow, only allowing the fine material of floods to be spread across the cave floor in a gentle sheet. By the time the earth finally settled to rest, the entrance to the large, airy cave—fit enough for a dragon—was raised high and dry above the normal flow levels of the river.

A cave 40 metres high and 107 metres wide, with a shaft of light and air coming in near the middle, close to water, with views down onto a wooded plain with meandering herds of animals was a magnet for humans who soon discovered and occupied it. The shaft opening would have created a small waterfall during periods of heavy rainfall. So, the people lived mainly in the eastern part of the cave, near the entrance, where the roof was well preserved. Here, they sat in small groups, communally knapping flakes from rocks, processing plant material and skins and huddling close to the fire during winter or when the northern glaciers crept southward.

Around 400 000 years ago, the eastern section of the cave roof collapsed, blocking the entrance to the cave. The people moved their activities to the western part of the cave where many recesses provided shelter. Here, thick ash layers developed—in places exceeding 10 metres—evidence of the constant presence of humans in the region for millennia.

By 230 000 years ago the cave was almost completely filled up with sediment and residue from 300 000 years of more or less continuous use. From the evidence it appears as if Peking Man lived in the cave for a more or less continuous period from at least 500 000 years ago when the cave floor rose above the level of the river, to 230 000 years ago. During this span of human occupation, the climate oscillated between colder and warmer several times. Evidence of these times has been trapped in the sediment of the cave showing two dry, cold periods, separated by a more temperate period. Also trapped in the sediment is a large and diverse collection of

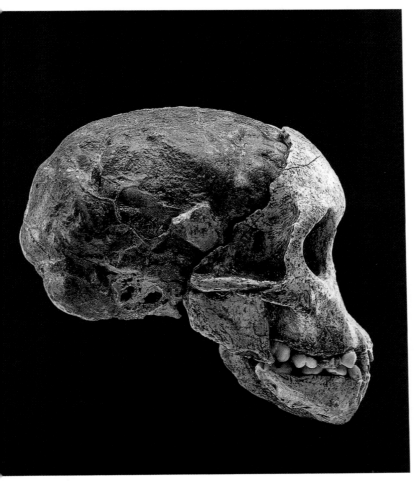

Skull of *Australopithecus africanus*, from Taung, South Africa. If the famous Taung child matured at a rate similar to modern apes, then its estimated age at death was three years. *(Photograph by David Brill)*

Overleaf: Frontal views of the Taung child skull. The first early hominid found in Africa, this skull provided the basis for a new genus and species. *(Photograph by David Brill)*

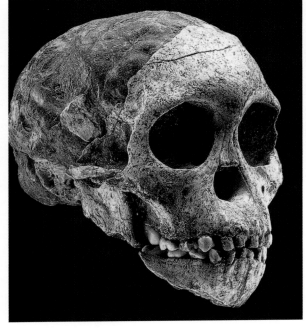

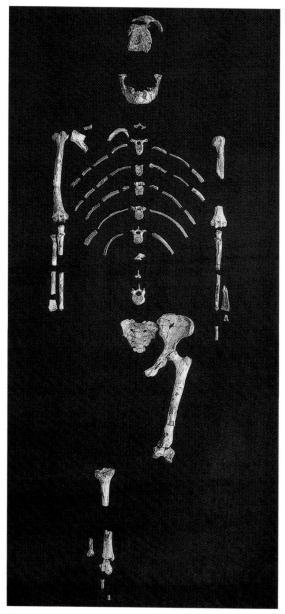

Skeleton of *Australopithecus afarensis* from Hadar in Ethiopia. More than two decades after its discovery, Lucy's skeleton remains an important reference point to which older and younger fossils are compared.

The site of Laetoli in Tanzania attracted worldwide attention after the 1978 discovery of a hominid footprint trail that demonstrated a humanlik gait for *Australopithecus afarensis*.

human remains—the largest in the world from a single site.

It was a frigid day in October 1927 with the temperature down to below zero and a hazy, pale sky, smudged with airborne sediment, promising another dust storm from the Gobi Desert in the north-west. Sixty men, clustered in small groups or strung out in lines, were working in a quarry on the north side of a small outlying foothill above, and to the north-west of the town of Chou K'ou Tien. Men were pounding at the rough limestone with picks and hammers, hacking away the hillside, helped occasionally by mild charges of dynamite. All the dislodged rock and rubble was loaded into baskets, slung one on each end of a pole, and carried by other labourers along sagging planks up the steep, winding path to the upper level of the hill, there to be examined by palaeontologist Birgir Bohlin.

Bohlin, together with Chinese geologist C. Li from the Chinese Geological Survey, had arrived in May 1927 to oversee the new dig at Chou K'ou Tien, for which the indomitable Davidson Black had secured funds from the Rockefeller Foundation. The plan for 1927 was to make a systematic survey of the site and to clean up the talus strewn about the ground; to blast off the dangerous overhanging breccia and to investigate the deposits brought down by this operation; and finally to start a systematic excavation of what remained *in situ* of the original cave deposit itself. It was the largest excavation ever undertaken in search of human ancestry.

The locals of Chou K'ou Tien, however, weren't caught up in the enthusiasm. Empowered by the growing communist movement, they sat down on the site and tried to prevent the work when the small army of diggers arrived at Chou K'ou Tien. The Chinese had a point. Foreigners had a long tradition of discovering—and confiscating—the cultural treasures of the Chinese. Names like Sven Hedin and Roy Chapman Andrews were famous in Europe for their daring trips into remote and dangerous regions. They were notorious in China for taking treasures out of the country. Hedin, in particular, was known as a discoverer of ancient ruins and long unused trade routes, and even claimed to have rediscovered the old Silk Road of Marco Polo, which he said had been forgotten by modern

Chinese. On one occasion the Chinese insisted on his taking some Chinese scientists with him on an expedition to Mongolia. Hedin overcame the 'difficulty' by dropping off the Chinese at intervals and putting them 'in charge of' meteorological 'stations'. This flattering honour pleased them and he proceeded without further supervision by the Chinese government men. For him and others of the era China was simply an exploratory playground.

Even the Swedish scientist Gunnar Andersson showed little inclination to share his fabulous cache, including everything from dinosaur eggs to exquisite dynastic pottery, with the Chinese. In 1921, for instance, at the end of one of the most severe famine periods in the modern history of China, Andersson journeyed by boat up the Yellow River to the province of Honan searching for graveyards of fossil elephants buried in the loess. At one of the small villages he sent his cook and his collector, Yao, to ask questions about likely sites. Later that evening the cook and Yao independently brought back fine specimens of elephant tusks. Early the next morning a trickle of villagers brought pieces of elephant's skull for which Andersson paid a bartered price. Then, like a dam-break, a continuous stream of villagers brought bits and pieces, and sometimes even baskets full, of elephant fossils. The word was out. Now eager to see the site, before all the fossils were plundered, Andersson followed a track, flanked by a host of curious onlookers, for 3 kilometres through a deep and narrow ravine. On approaching the site he was met by the actual owner of the property. Perhaps not surprisingly, the owner was not happy with Andersson's celebrity presence, and the chaos caused by it.

Like everyone else, for three years in succession the owner's wheat crops had failed. Then, just when a fine harvest was almost ready for gathering, hordes of villagers trampled down the swathes of his narrow, painstakingly cultivated, terraced strips in order to reach the side of the ravine where the fossils were located. Andersson, having sympathetically noted the destructive result of his enquiries, nevertheless continued making his way to the site, together with a crowd of 200 others, massively contributing to the destruction. The owner, amazingly still polite, was reluctant about letting Andersson continue digging out what was left of the fossil elephants. But Andersson proceeded to dig, without further negotiation, proposing to pay according to results. Just as he was comfortably seated in the elephant cavity, he heard a shrill and angry woman's voice. It was the aged mother of the owner, and her clamorous complaints related to the trampled wheat. Andersson relates:

I had long since learnt that in China everything can be put right except hysterical old women. I realised that the situation was very serious, so I asked Yao to propitiate the old lady. Yao could talk quite charmingly when he really did his best, but this was of no avail, for the old woman worked herself up into an even greater frenzy. Finally she climbed up resolutely into the cavity, in this way preventing all further excavation work. It was undoubtedly most amusingly done and was a genuine sample of the famous passive resistance of the Chinese.

I opened my parasol and held it over the old woman, an action which gave rise to some merriment, but did not materially alter the situation. I knew there was one infallible means of driving out a Chinese peasant woman, so I made preparations to take a photograph of the spot, as I wished in any case to have one. I was right; the camera had the desired effect: the old lady got up from the hole and crept behind the back of her son.

It seemed that Andersson had finally caused action-able offence, and Yao, noting the changed mood of the crowd, advised him that if they didn't leave immediately they would be in big trouble. Andersson concluded, 'I sincerely hope that until the day of her death the old lady may, as a slight com-pensation for the damage to her trampled wheat, enjoy the great honour as one who cast out a for-eign devil'.

Luckily, Davidson Black was from a different mould and he felt that the Chinese should have more say in what specimens, if any, were to be removed from their country. Indeed, it was on this basis that he established joint research into the antiquity of Man in China. Black eschewed colo-nialist policies when it came to research, and it was probably the footing of equality that he established between the Chinese and the Peking Union Medical College that made him feel secure in what was clearly a very insecure situation.

Throughout 1926 and into 1927, the increasingly shaky Nationalist Party and Communist Party coalition moved northward in a United (by name only) Front. This 'northern expedition' aimed to destroy the northern warlords and ultimately reunify China under one rule, that of Chiang Kai-shek.

In the cities to the south, which had fallen to Chiang Kai-shek, there were terrible rumours of massacre and bloodshed. These horror stories sent shock waves to Peking. In November 1926, Mrs Fortuyn lost her nerve. She took her three children out of China via the Trans-Siberian Railway, leaving her husband to slog it out at the Peking Union Medical College. There, the nervous tension increased to breaking-point as Fortuyn noted in his diary entry of 27 February 1927:

> Although Peking is supposed to be an open city, the danger of insurrection of the masses may not be unthinkable. The fear of nervous breakdown is much greater. Recently a young American doctor precipitately returned to the United States. Not everyone can stand the peculiar tension which goes with life in Peking.

In March 1927 Nanking fell to the Nationalist and Communist forces. Communist soldiers went on a murderous rampage venting their anger at the despised foreign element in the city. Chiang Kai-shek, looking for any excuse to come down on the communists, countered with a massacre of his own. Anyone who was communist, or even vaguely associated with them, was slaughtered. Meanwhile some of the northern warlords were making a push for Nanking from the north, shelling the city from across the Yangtze River which became a river of blood of tens of thousands of troops.

Fortuyn's diary entries mirror the events:

March 29, 1927: The situation of the Peking Union Medical College begins to deteriorate. The Dean and Director send away their families. Men stay or return after taking their families to safety. The political situation is tense. All the Chinese are calm, but daily expect Peking to be in the hands of the Southern Army [Chiang Kai-shek's].

April 2, 1927: Mrs Black and the children leave for America. The Rockefeller Foundation insures private property against all risk without cost to us—one less anxiety.

April 7, 1927: I am depressed—but it will pass.

April 9:, 1927: There is a noticeable relaxation—not so much because serious things do not occur anymore as because so much is being accepted without reaching the breaking-point.

As 1927 progressed, the front-line of fighting between the northern warlords and the southern armies of Chiang Kai-shek approached Peking. Meanwhile the Communists, who had been routed by Chiang Kai-shek, took their battle underground and into the countryside.

In the Peking Union Medical College, Davidson Black's reactions to the 'unsettling conditions', was to comment that it quite disorganised his correspondence, and forced him to put his three years' work on prehistoric human skeletal material in order. 'Don't take it too seriously' was his motto.

At Chou K'ou Tien, in the middle of the war zone, a revolution of another kind was occurring. Here, after the dispute with the locals had been resolved, the workers were tearing off a whole hillside from the mountain range. During the long, hot summer of 1927 they laid bare the cavities of Dragon Bone

Hill east and west for a distance of five metres with a breadth of 16 metres north and south. Even though only part of this was fossil bearing, all material taken out was inspected. The greater chunks were transported to Peking and the rest carefully sifted and examined locally. Summer drifted into the fine, chilly weather of autumn. Soon it would be too cold to work. But still, against expectations, nothing spectacular was found.

It was mid-October and only three days to go before the first year's systematic investigation of Chou K'ou Tien would end. With warring armies north and south, and a welter of geological fragments and general confusion, palaeontologist Bohlin remained intent on his work. Then, miraculously, like a needle in a haystack, he spotted it—a perfectly preserved human tooth, emerging from a lump of limestone.

On 19 October 1927, Bohlin unexpectedly strode into Black's office in dusty field dress, beaming from ear to ear, delivering his treasure, a beautifully preserved human left lower molar, by hand. Black

described the scene to Gunnar Andersson in a letter which provides a fascinating insight into the circumstances:

We have got a beautiful *human* tooth at last!

It is truly glorious news, is it not!

Bohlin is a splendid and enthusiastic fellow who refused to allow local difficulties and military crises to affect his work. I was telegraphed on October the 10th urging me to recall Bohlin and Li because of the war but I made careful enquiries here and could see no reason for so doing. So I wrote to Bohlin and explained the situation and told him I didn't want him or Li to run any risks but that as they knew local conditions they must use their own judgement about coming to Peking.

They had 60 men on the job so their daily payroll was relatively large. I couldn't get away myself for I was having daily committee work that demanded my presence here. At any rate I couldn't reach Chou K'ou Tien on account of local fighting. That night which was October 19th when I got back to my office at half past six in the evening, I found Bohlin in his field clothes, covered with dust, but his face shining with happiness. He had finished the season's work in spite of the war and on October 16th had discovered the tooth; being right on the spot when it was picked out of the matrix! My word, I was excited and elated! Bohlin came here before he had even let his wife know that he was in Peking—he is indeed a man after my own heart.

To Black, the tooth was worth more than all the gold in China. In 1927, on the evidence of the new tooth which may have been as old as half a million years, and the two earlier finds, Black boldly gave the new human the generic name *Sinanthropus pekinensis* (the Chinese man from Peking) a new genus of human being on Earth, perhaps our ancestor. Needing to take the tooth overseas to confirm the validity of his new classification, Black wore it around his neck in a specially made brass capsule with a screw closure and a ring at the top through which a strong ribbon was threaded. As happy as a little boy, he would protect it with his life.

Black travelled widely in 1928 with his one small piece of evidence of early man in its capsule. He visited the Rockefeller Foundation to discuss his research program and to ask for top-up funding to keep the Chou K'ou Tien work going until March 1929. He also visited European and American scholars seeking confirmation of the validity of the establishment of his new genus. While some scholars were in general agreement with his conclusion, others received the classification coolly, feeling he was premature in his belief that he had found evidence of primates in China. One small fossilised molar was, surely, not enough on which to base such a momentous claim. This had little effect on Black other than to make him redouble his efforts to establish the proof of his claim.

Back in China things were looking grim. The excavation site for 1928 lay to the east of the one selected in 1927, and the digging was begun from the top of the hillock. But no sooner had the work started than the war began to spread into the Chou K'ou Tien area. Fortuyn, who was left in charge while Black was away overseas, recalled the team in May. They were able to return to Dragon Bone Hill only in late August. To make up for lost time the team stayed in the field until 25 November when the winter's snow and ice finally drove them out. The catch for the year—the right half of the lower jaw of *Sinanthropus* with three permanent molars *in*

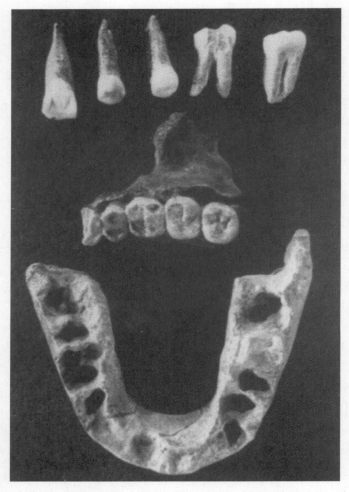

Some of the remains of Peking Man: Top: teeth; middle: maxilla; below: mandible.

situ—was found once again in what was becoming a magical time, just days before the end of the season.

Among the 2800 cubic metres of deposits and 575 boxes of material, the finds weren't much, but they were enough for Black to courageously ask for yet more funding from the Rockefeller Foundation. He wrote a detailed memorandum documenting the conditions in China and explaining the importance of keeping research active in order not to lose the success of the past two years. He pointed out how fortunate it was that the workers at Chou K'ou Tien had not been seriously molested by what he termed reactionary forces. There was no knowing, however, what would happen if the site was abandoned. The importance of keeping the trained men at work was paramount to prevent them being forced into the army. The strong group of professional workers, both Chinese and Western, was invaluable and might never again be available should the present situation make it difficult for the Chinese and foreigners to work together. He also described the desperate financial situation of the Chinese professionals, pointing out that the Geological Survey would collapse without financial aid. Black was determined to keep his co-operative endeavour a going concern.

To forestall impending disaster and to advance the work, Black proposed the creation of a Cenozoic Research Laboratory. It would be a special department of the Chinese Geological Survey for the investigation of Cenozoic era (the past 60 million years) geology and palaeontology. All the material investigated would remain in China. If such an organisation came into being, Black said, it would, literally and metaphorically, break new ground in China. It was obvious that such a plan could not be carried out without the substantial grant from the Rockefeller Foundation.

The reply came on 5 April 1929. Fortuyn recorded it in his diary: 'Today the news came that Dr. Black received in US funds $80 000 from the Rockefeller Foundation for his research for human fossils'. In an understatement, he added 'quite a bit!'.

The field season for 1929 was drawing to a close. Davidson Black had made frequent trips to Chou K'ou Tien, travelling the first 20 kilometres by train (if it was running) or by automobile in a Model T Ford, along tracks not much wider than a donkey trail. Often it was faster to go the last 30 kilometres by rickshaw pushed by sinewy peasants, or on the back of a donkey. The trip took an entire day.

Black's enthusiasm powered the whole enterprise as excitement was beginning to flag. The team had been searching hard and long for three years now, for larger parts of *Sinanthropus*, hoping ultimately to find a skull. The Cenozoic Research Laboratory's reputation was on the line and depended on something more substantial than a tooth and fragments of jaw bones. News of more finds was eagerly awaited by scientists in Europe and America, many of whom had taken the long journey to Peking. By late 1929, the number of visitors had fallen off and interest was dwindling. There remained in Peking only a few staunch scientists and the remotely curious international population.

Chinese technician Pei Wenzhong was now in charge of the field component of the Dragon Bone Hill excavation, replacing Birgir Bohlin and C. Li who had gone on expeditions into remoter regions of China. Pei had arrived in Chou K'ou Tien in April to discuss the 1929 excavation plans with Davidson

Scientists in front of a camel caravanserai where they stayed as a study group near Chou K'ou Tien. Left to right: Pei Wenzhong, Wang Hengsheng, Wang Gongmu, Yang Zhongjian, Birger Bohlin, Davidson Black, Teilhard de Chardin and G. B. Barbour.

Black. Other deposits had been found over the past several years, but Black decided to stick to the main deposit which had come to be known as Locality 1. It had originally been a great cave that had gradually been filled in through weathering action, periodic roof collapse and the accumulation of detritus. Over time, water seeped through the accumulated scree and detritus, depositing lime which cemented the fragments into a hardened pudding of rock types, called breccia.

During the field seasons of 1927 and 1928 some 6000 cubic metres of the cave deposit had been excavated, more than 1000 boxes of fossil material had been transported to Peking for preparation and, above the level of the quarry terrace, the deposit had been exposed over a vertical face some 18 metres in height by 28 metres in width. And still there was no evidence that the deposit was being exhausted, with the lower limit not having been reached.

Four distinct layers had been hacked through, each representing a stage in the life of the cave. Layers one and two were 4.5 metres in depth, and contained fossils of bats and blocks of what had been stalagmites, evidence that even the top of the hill had at one time been a cave itself with the superstructure now eroded away. Layer three included a trail of huge limestone blocks, evidently representing a collapsed section of the roof. It was 3 metres deep. Layer four was a thick, 5.5 metre accumulation of thinly laminated coloured clay containing abundant fossils of small animals and some bones of large animals. Tellingly, most of these appeared burnt, as if barbecued. In addition, there were numerous stone artefacts in this layer including quartz fragments, clearly brought in from elsewhere.

Pei and Black, discussing these layers, decided that the target would be the middle section of the areas that had been dug in 1927 and 1928. Pei would start digging at the fifth layer, which had mistakenly been identified as the bottom because it was, literally, rock hard. This layer was actually a slab of black fossiliferous clay embedded with calcium. Pei had no idea how thick it was.

After Black left, Pei's small army of workers tried day after day to break through the layer. Even dynamite would not budge it. It didn't help that he was living alone in a small temple in the mountains and had the sole responsibility of supervising the most important dig in the world. Under these circumstances it was not only the rocks that were feeling the stress.

Pei's persistence paid off, however, and the fifth layer was finally broken through after three weeks of battering. Though hard, it was not thick, and at the sixth layer, more fossils were found. As the digging continued down eight metres into a sandy seventh layer, a rich assemblage of exceptionally well-preserved fossils was unlocked from its stony matrix including 145 jawbones of thick-jawed deer, found in a single day. Complete pig, buffalo skulls and deer antlers were also recovered from this layer. It was a pity that the work area, now easy to dig, was now also restricted due to dangerous overhanging slabs of rock that threatened to slide down on both the southern and eastern sides of the dig.

By the end of September, the excavated area gradually narrowed to a space big enough for only a few men to work in. Pei thought that they had finally reached the bottom, but then he sighted a crack in the southern side, about 40 metres deep. He and another worker were precariously lowered into the crack, fastened by ropes at their waists. As if in a

Above and facing: Digging in process at the point where the main cave and the north fissure coalesce. This is the spot where the Peking Man skullcap was found.

dragon's lair, they were delighted to discover a treasure of fossils.

It was now the end of November and fieldwork was meant to be suspended for the season; but, as if weaving a spell, the many fossils in the narrow chasm hypnotised the men to continue working for a few days more. It was, once again, that magical time at the end of the season.

It was four o'clock in the afternoon on 2 December 1929, near sunset with a winter wind bringing freezing temperatures to the site. Everybody felt cold. There were four men down in the chasm, working in a space so narrow that each had to hold a candle in one hand and work with the other. Maybe because of the cold weather, or the hour of the day, the stillness of the air was punctuated only by the occasional rap of a hammer indicating the presence of men down in the crack. 'What's that?' Pei cried out as a feeble candle-flame of one of his companions flickered over a curved shape. Pei scrambled over to the spot, holding his breath in excitement. 'A human skull!' In the tranquillity, everybody heard him.

A rounded shape of a skull was, indeed, emerging from the matrix of rock. Half of the skull was freed, but half was still embedded in rock. Some workers suggested that they might damage the object if they continued to work by candle-light after dark. There was some logic in this as one labourer pointed out. 'It has been there for so many thousands of years, what harm would it do lying there for one more night?'. But a long night of suspense was like a million years for Pei Wenzhong. He decided to get it out then and there.

The skullcap of a female, before she was carefully removed from her millennia-old resting place.

The earth, however, would not give up its child so easily and Pei needed to extract a large, heavy chunk of rock which embraced both the skull, embedded in a large adherent block of cemented matrix, and a second block containing a small fragment of the specimen. Pei recounted that he was too excited to eat supper:

I was thinking how to let the Peking office know. I wrote a letter to Weng Wenhao [the head of the Geological Survey of China] and dispatched a man to deliver it the next morning. After the messenger had gone, I felt uneasy again for the letter would reach him only late that night when everybody was in bed and Weng would have no time to let the people concerned know about the big event. So thinking, I sent a telegram to Black saying: 'Found skullcap—perfect—looks like man's'.

The fossil itself was rather wet and soft and easily damaged. Pei took it immediately to a fireside in the field headquarters—spartan adobe rooms in a local caravanserai—and stayed a day and a night to dry the specimen. Then he wrapped the precious burden with infinite care within layers of Chinese cotton paper and then with a heavy layer of coarse cloth impregnated with flour paste. The weather was so cold that these plaster wrappings did not dry even after three days in a comparatively warm room. Since he was afraid to transport such fragile specimens in wet wrappings the blocks were further dried on the night of 5 December with the aid of three braziers. He then wrapped the lot in two thick cotton quilts and two blankets and tied it with rope like ordinary luggage. Pei then put it in the basket of his bicycle and rode to the nearest bus stop. A bus trip to Peking with a strange bundle was no easy matter in a suspicious and strife-torn China and Pei took the trip with trepidation. He was prepared for the worst as his own account explains:

Then we travelled by bus between Fangshan and Peking. There was a check-point at the Xibianmen Gate in Peking where luggage would be examined as a matter of routine. I made preparations for that with a few fossils to show to the officer, intending to tell him it was the same kind of thing inside the wrapped luggage and to ask him not to open it. If he insisted on opening the bundle, the plaster and gauze would have to be kept intact. If he still insisted on taking a look at what was inside, I would ask him to arrest me first. The man was polite. I only had to open the bundle to let him see.

Black's excitement was beyond description. Here, finally, was concrete proof—indeed proof locked in concrete—of the existence of another species of human. He shot off letters to his closest friends and colleagues reporting on the glad tidings and praising the scientists doing the fieldwork at Chou K'ou Tien. In one letter he wrote:

> Again the year's work at Chou K'ou Tien closes with a grand climax, for on 2 December W.C. Pei in charge of fieldwork there discovered the greater part of an uncrushed adult *Sinanthropus* skull! He recognised it *in situ* and excavated it with care himself bringing it to me on 6 December with the field wrappings still wet. The whole mass included a great block of travertine in which the skull is half embedded...Mr Pei is a corking field man!

On 28 December a special conference of the Geological Survey of China was held for an audience of scientists and journalists. News of the extraordinary find flashed around the world and it became the topic of the moment. Black was catapulted to fame. He was not, however, ambitious for himself and rather than basking in glory, he himself took on the delicate task of freeing the skull from its shrouding of rock. Relaxed and in good spirits he wrote to a friend:

Yes, *Sinanthropus* is growing like a bally weed. I never realised how great an advertising medium primitive man (or woman) was till this skull turned up. Now everybody is crowding round to gaze that can get the least

Sinanthropus pekinensis.

excuse to do so and it gets embarrassing at times. Being front page stuff is a new sensation and encourages a guarded manner of speech! The work of unmasking the villain or of extracting *Sinanthropus* from his clinging hard matrix progresses slowly and I am now trying to devise a vacuum cleaner effect to keep from premature silicosis. I wish you would see the halo of dust after fifteen minutes of grinding with a dental carborundum point under an airblast—talcum or flour isn't in it with this.

Not only couldn't he breathe while he carefully drilled away the hard rock, but worse, his spectacles were so thickly covered that he couldn't see. His wife complained that he arrived home every night looking as if he had spent the day in a coal mine. She couldn't recognise the spectre that was Black when he came in the door.

By April 1930, the team was again off to Dragon Bone Hill, eager to find what lay beneath the skullcap. This time, however, they went to work with a feeling of security they had never had before, since the site had been purchased by the Geological Survey to be preserved for science.

Points and scrapers fashioned and used by Peking Man.

As the discoveries multiplied so did Black's honour and his fame. In 1932, he was awarded Fellowship of the Royal Society, one of the most prestigious awards in the world for a scientist. During the latter part of 1932 and most of 1933, he circled the world giving lectures and meeting with scientists. Much as Black enjoyed these meetings and other more informal affairs, where he could discuss his work with his colleagues and friends, they all took their toll of his time and kept him away from China and from producing the definitive report on the history of the discovery and identification of *Sinanthropus*.

In the autumn of 1933, an exhausted Davidson Black returned to China. His colleagues noticed that his pace was slow and he lacked his usual boundless energy. Nevertheless, no sooner had he settled in than he was off to Chou K'ou Tien to catch up with the excavations. By 1933, the tally of human material found at Dragon Bone Hill was a number of teeth, several jaws, two complete and several fragmentary skulls, traces of fire (charred bones and layers of blackened deposits like ash) and myriad stone tools.

It was the many quartz fragments that, in 1918, had sent Gunnar Andersson's imagination spinning toward the conclusion that Dragon Bone Hill contained the remains of ancestral Man. In what seemed to be a blind spot, the common occurrence of such fragments, ten years later, still did not attract much attention. Indeed, in 1929 Black stated, curiously, that the site had not yielded any artefacts of any nature. When Pei Wenzhong made his momentous discovery he also reported the recovery of one piece of quartz at the site that showed marks of a 'blow'. He couldn't say, however, whether the piece had occurred in association with the skull. Then, in 1931, a layer of loose deposits was found that was loaded with quartz fragments, enough to fill large basketfuls of stone artefacts daily.

Such a haul of such antiquity had never been seen before. It was enough to draw the French Professor, Abbé Breuil, the world's pre-eminent expert on stone tools of the early Stone Age, to Chou K'ou Tien. Fascinated, he sat down at the site and literally began banging stones together to ascertain the skills of Peking Man. Using pebbles from the river bank as hammers, Peking Man knapped flakes from chunks of quartz, quartzite and chert. Some of the quartz was from a granite hillside 2 kilometres away. Among the types of tools found were choppers made of round, flat sandstone or quartz pebbles, made by trimming one or both sides. These were probably used to fell trees or to make hunting sticks. Scrapers of all sizes were used as general purpose knives to carry out a number of activities such as skinning animals, separating the sinews from the bones or digging out larvae from under the bark of trees.

The apparent use of fire, too, was a landmark in human history. At the beginning of the excavation no-one knew what the layers of black substance were. Then, in 1930, chemical analysis revealed that it was a residue of carbon. Together with charred bones and rocks, the evidence for fire became virtually indisputable. *Sinanthropus* may have been Stone Age but she was no block head.

When Black reached his beloved *Sinanthropus* site in 1933, he suffered a transient heart attack but recovered quickly and went on with his inspection. Back in Peking he took a medical examination, the results of which were very grave. Except for his secretary, who was sworn to secrecy, he told no-one, including his wife. After a brief rest, he resumed work and began to put his affairs into order, with his usual habit of working through the night. On 15 March, Davidson Black went into his laboratory at five in the afternoon. A young Chinese colleague went to pay him a visit and found him sitting at the desk where he had worked for years and years at science. Black spoke anxiously about the future of the research in the rapidly deteriorating political situation. His visitor left and the quietness of the evening came to the College. Black turned once more to his work. An hour later he was dead.

4
Missing
in action

In the early part of the seventeenth century, an avaricious Britain was using every means to open a reluctant China to trade. Ships carried silks, dried rhubarb (favoured as a laxative) and Chinese tea to England. The Chinese, however, wanted little in return except Sterling. That was until the British initiated the opium trade.

The British imported the opium from Turkey or India and encouraged its addiction. In 1800 the Chinese imperial government reacted to the degradation of its people and the consequent outflow of Sterling by forbidding the import of opium. The restriction fell on deaf ears and, if anything, the opium trade flourished. Indeed contraband shipments—40 000 bales of raw opium entered China in 1838 alone—were worth twice as much as all legal trade.

In exasperation, in 1839, Chinese authorities at Canton confiscated and burned 20 000 bales of opium. Ruthlessly taking advantage of China's extreme vulnerability to sea power and lack of modern technology—steamships, rifled cannon and exploding shells were so fantastic and alien to the Chinese that they attributed them to magic—Britain used this action to provoke the ensuing Opium War from 1839 to 1842, humiliatingly defeating China. This inaugurated a period of unequal treaties (which lasted in law until 1943), during which China was carved up among various foreign countries like a ripe melon, losing her autonomy and being forced to submit to the creation of numerous foreign-controlled enclaves, or Treaty Ports. Japan's influence was mainly in the northeast provinces of China—Liaoning, Kirin and Heilungkiang—collectively called Manchuria.

From the 1880s the most visible sign of the modernisation of China was the construction of railways.

In Manchuria, the Southern Manchurian Railway Company was owned by the Japanese who received considerable concessions around the area of the railway line, including police powers along the right of way.

By the First World War, Japan was second only to Britain as a foreign investor. The War had stimulated the textile industry worldwide and Japan, using China's cheap resources and wielding the economic whip, was the main beneficiary. Since postwar Britain had domestic problems, Japan soon became the Great Power. She had the largest vested interest in maintaining the existing system of unequal treaties, ironically using the 'gunboat diplomacy' tactics that Britain had initiated during the Opium Wars. Alone among the foreign powers who toyed with China, Japan was in good shape, having not been depleted by the First World War. Mercilessly, Japan used her large and capable conscript army to bully China into accepting a number of crippling demands, which included extension of Japanese leases in southern Manchuria, and 50 per cent ownership of major State enterprises including iron and steel complexes.

Influential warlords were also on Japan's payroll including, for a time, even the leader of the so-called Chinese Government. As Japan's influence grew, so did Chinese hatred for the Japanese. Strikes, boycotts and acts of violence against Japanese enclaves increased. In return, the Japanese became increasingly impatient with the Chinese Nationalist movement.

In 1931, this irritation boiled over and the Japanese blew up a section of the Southern Manchurian Railway—their own railway—blaming it on the Chinese. Using this as a pretext for protecting their interests, the Japanese invaded Manchuria, cynically proclaiming it an independent country by installing a puppet emperor, and naming it Manchukuo, the country of the Manchus. Not content with this, the Japanese then seized the strategically important main railway junction north of Peking. A move on the adjacent province of Jehol and even the capital itself seemed imminent.

The United States' reaction was mild. In 1932, it officially created a policy of non-recognition of Manchukuo. The League of Nations (all those who had interest in China, including Japan) appointed an investigative commission headed by the Earl of

Lytton, a British diplomat. The Japanese delegates, being aware that the Lytton report would condemn them, stormed from the League of Nations and Japan swiftly seized Jehol, thus shoring up its power-base. While it may have been a game to the foreign governments involved, in China it looked like a setting for another World War. It was March 1933. No wonder Davidson Black was so anxious the hour before his death.

In reality, the United States and League of Nations would not fight Japan—at least not over China. A truce was signed in May 1933 that allowed Japan to keep Jehol, as well as maintaining power over Manchukuo.

Chiang Kai-Shek, ignoring Japanese aggression, was, in the meantime, hell-bent on suppressing the Chinese communist movement which he had all but pulverised in five separate campaigns, sending the stragglers on their Long March in 1935. The Japanese, he said, were merely an irritation of the skin while the Communists were a disease of the heart. Despite increasing pressure on him to wage war against Japan, Chiang realised that he needed time to radically modernise his forces before his Nationalists could hope to win a war against them.

In fact, he had the nagging suspicion that no amount of time would permit China to defeat Japan without foreign help. It was this doubt that had led him, through the years, to pursue a policy of appeasement; a policy that was becoming increasingly unpopular.

There would be no rest for the Communists, however, and in 1936 Chiang flew to Sian in central China to organise one last campaign to destroy them in their boltholes. Under extraordinary circumstances, however, Chiang was kidnapped by a local warlord who was playing his own power games. Retaliatory proposals for an armed expedition against Sian were overruled by Nationalist leaders, including Madame Chiang, in favour of negotiations for the Generalissimo's release. In what was an extremely adroit tactic, the Communists, backed by a Japan-loathing Russia, sent a high-level diplomatic delegation to work for the release of Chiang. Of course there was a price attached for Chiang: he was to bury the hatchet and allow a Second United Front of combined Nationalists and Communists to be established for the specific purpose of fighting the Japanese. The Communists were successful in the negotiations to release Chiang Kai-Shek, and the apparent reconciliation between the Nationalists and the Communists was wildly popular. Except in Japan.

After Davidson Black's death, Dr Fortuyn took on the position as head of the Anatomy Department of the Peking Union Medical College, a position he had, at any rate, been acting in for some time. The new head of the Cenozoic Research Laboratory (combining the Geological Survey of China and the Anatomy Department of the Union) was Dr Franz Weidenreich, who arrived in Peking in April 1935. He was to head the Chou K'ou Tien dig.

The dig at Dragon Bone Hill had continued in the two seasons since Black's death. But, as if the whole site had gone into mourning, only 'commonplace' objects were found: bones of 37 different types of mammals including bears, hyenas, sabre-toothed tigers, horse and a variety of deer and other herd animals; extensive evidence of the use of fire including thick deposits of ash; and many stone tools. No more human remains were found.

Toward the end of 1936, tension at the site was great. Money was running out and so was hope for

Jia Lanpo at the Peking Man site, 1936.

more significant finds. Weidenreich didn't help. He obsessively controlled expenditure to the point that piles of soil dug up by the excavation team were left unsifted (in order to save money), which only added to the stress. Weidenreich visited frequently, by now being able to do the return journey to Chou K'ou Tien from Peking in a day by car. His visits were characteristically short and abrupt. 'Anything new? A skull may be lying below! Sorry I can't stay long. I have to go back to the city this afternoon.'

Finally, on 22 October 1936, a complete lower jaw of Peking Man emerged. It had been years since anything substantial had been discovered and the team was ecstatic. The leader of the dig was now Jia Lanpo. In a re-run of the 1929 find, which made the site world-famous, the skies were overcast and the weather freezing. Jia nevertheless decided to go on with the digging. It was 15 November, a Sunday. There had been a heavy snowfall the night before and the team waited until 9.00 a.m. to start digging. Not many minutes later, Jia noticed a technician routinely putting a fragmentary bone into a wicker basket. The shape of the piece looked like a human skullcap. Jia rushed over, took a second look and sure enough discovered part of a human skull embedded in some rock.

Jia immediately marked out a 6 square metre zone and selected the ablest men to carefully work with him in that zone. In an area of half a square metre they found not only fragments of a skull but also bones to make up part of the face. Then, unexpectedly, in a deposit half a metre below and 1 metre away from the first skull another turned up. Two skulls in one day!

Jia humorously recounts Weidenreich's reaction:

Not until Monday morning did the telephone message from Chou K'ou Tien reach Weidenreich, who, as related by his wife, got so excited that he put his pants on inside out. That afternoon when he came to Chou K'ou Tien and held the first skull, which it had taken four people the whole night to glue together, his hands were shaking, and he uttered the words 'Wonderful! Wonderful!'.

But this wasn't the end. On 26 November, long after digging should have halted, another skull was unearthed. Better preserved than any of the previous skulls it was taken to the city by Jia himself, while the site froze over for the winter. Digging

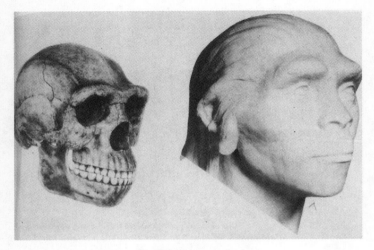

Female *Sinanthropus pekinensis.*

Male *Sinanthropus pekinensis.*

resumed in April 1937, with the onset of spring and a new flow of funds from the Rockefeller Foundation. The site continued to offer up its buried treasures in a cathartic burst. By the end of May, several teeth, some skull fragments, a skullcap, a brow ridge, a lower jaw and, for the first time, an upper jaw were unearthed. The upper jaw allowed Peking Man, finally, the dignity of a whole face, ironically at the time when the rest of China was about to lose face to the Japanese.

The last pre-war excavation occurred at Dragon Bone Hill on 9 July, two days after the break-out of the Sino-Japanese War in 1937–41, which started after a minor incident at the nearby Marco Polo Bridge. Here, Nationalist and Japanese troops had a minor skirmish in which one Japanese soldier went missing. For a few days it appeared as if the incident could be settled by a local truce between the opposing armies. Chiang Kai-shek, however, used the incident to publicly denounce Japanese aggression, saying that no more Chinese territory would be alienated to Japanese control. The bellicose

Japanese, however, promptly announced the annexation of the provinces of Hopei and Chahar, which were adjacent to Japanese-controlled Manchuria (Manchukuo) and Jehol. Japan now controlled all of northern China.

Not that the Chinese military could do much about it. Despite Chiang Kai-shek's efforts, they continued to play warlord politics. The result was hopeless disorganisation. This endless organisational variety meant that training and maintenance of any standards were impossible. In such chaotic circumstances, the fact that Japanese commanders could at least be expected to be obeyed gave them great advantage.

The field work at Chou K'ou Tien, being near the battle zone, was suspended. While some technicians went to Peking, 26 local men stayed behind. There had been a tradition that, during the off-season, some of these workers would keep an eye on the site, make some trial diggings and sift the earth again for overlooked fossils.

A 54-year-old technician, Zhao Wanhua, was heading the team that remained at the site, corresponding regularly with Jia Lanpo in Peking. Chou K'ou Tien soon fell under Japanese occupation, while

Staff and labourers at the Peking Man site. Jia Lanpo is first from the right in the back row. Second from the left and first from the right in the front row are Dong Zhongyuan and Zhao Wanhua, who were killed by the Japanese occupation troops at Chou K'ou Tien.

the hills around the little town were held by Chinese Communist troops. Jia, time and again, urged Zhao and the team to abandon the site and transfer to the city in separate groups so as not to attract the attention of the Japanese troops. Zhao finally took heed of Jia's pleading and began transferring members to Peking in groups of three from September 1937.

By the beginning of 1938 there were only three workers left, including Zhao, none of whom wanted to leave the site unattended. In the eight years that Jia Lanpo had known the men, they had been totally dedicated, never shirking a day's work. He was never to see them again.

The three field workers had been ordered by the Communists to serve in their kitchen. After several fierce encounters between the Japanese and the Communists the Communists retreated. The three men were captured by the Japanese who hoped to learn the identities and whereabouts of the Communists. The captives were tortured and finally, together with 30 others, were bayoneted to death.

The Japanese took Peking on 31 July 1939 in an uncharacteristically bloodless fashion (they had slaughtered the Chinese populations of the nearby towns of Tientsin and Tungchow). The Japanese then advanced along the major railway lines, north-west to Inner Mongolia, south to Hankow and to the Yangtze. Japanese air supremacy created panic: villages, even provincial capitals, would empty at the sound of propellers. It was a useful tactic and the Japanese thought it amusing.

The Japanese were succumbing to 'victory disease'. Mostly the symptoms manifested themselves in a form of brutal contempt for the Chinese population, expressed in rape, robbery, torture and casual killing, as an individual's whim or as policy of senior commanders. In battle, the disease took the form of over-confidence: after a series of easy scores the Japanese became cocky. On very rare occasions this worked for the Chinese.

One Japanese division became particularly careless and contemptuous after pushing their way through many easily defensible positions while invading one of the provinces of central China. The division began a march through a narrow pass, which had been worn through loess by centuries of ox-carts, without scouting ahead or securing the heights on either side. A division of Communists attacked the Japanese column at the head, centre and rear. The battle consisted chiefly of Chinese troops shooting and throwing hand grenades into the packed mass of Japanese soldiers trapped in the defile about 5 metres below. About 3000 Japanese were killed as against 400 Chinese casualties. It was rare that the Chinese got off so lightly.

By mid-1939 the Japanese had conquered all the foreign Treaty Ports of China, and Japanese rule stretched along most of the railway lines. The Japanese were unchallengeable at sea and had captured all of China's seaports, including Nanking. where the invaders carried out a well-publicised bloodbath—the Rape of Nanking, with the mutilation and slaughter of 300 000 Chinese civilians.

Chiang Kai-Shek, in the meantime, had removed his government to the remote Szechwan province, setting up in its capital Chungking. Szechwan and other remote central and south-western provinces of China were still held by the Chinese. In a desperate attempt to stop the Japanese from taking the railway south of the Yellow River, Chiang blew up the Yellow River dykes. It was an historic precedent—an event that had only ever been mirrored by catastrophic natural disasters. The river, one of the greatest on earth, shifted its course, flooding three provinces. Millions of people were drowned and millions more died of starvation in the aftermath. Peasants were cheap. In 1941 the Japanese took the railway anyway.

The Chinese population's hatred of the Japanese, and backlash against Chiang Kai-Shek's Nationalists, was manifest in the burgeoning Communist movement. The Japanese tried to reduce this trend by introducing a new doctrine in the summer of 1941. It was called 'Three All' meaning kill all, burn all and destroy all in Communist base areas. The statistics say it all—the population in the Communist strongholds dropped from 44 million to 25 million over a couple of years.

Chiang Kai-shek had been right about one thing. China's ability to wage war on Japan was limited at best and only foreign intervention would rid the country of the Japanese (and let the Chinese get on with their civil war).

By 1941, however, the Japanese had other priorities. It seemed they wanted nothing less than the world. Japan took advantage of the war with China (which was euphemistically termed the China 'Incident') to test her strength against Russia. Japanese forces took on the Russian army at the border of Manchuria and were humiliated. Not that this put a dint in the arrogance of the Japanese High Command. It only made them realise that to satisfy their voracious appetites, they needed to scale down the China Incident in order to build up their military strength. They turned their attention, instead, to South-East Asia, and occupied southern Vietnam in July 1941.

The occupation of Vietnam and the fact that Japan clearly sided with Germany in the Second World War, led the United States to place an embargo on all sales to Japan and to freeze Japanese assets. In what must be counted as the most severe case of 'victory disease' Japan retaliated by bombing Pearl Harbor on 7 December 1941, thus declaring war on the United States of America.

One of the first casualties of the Second World War was Peking Man. For a half a million years or more, remains of *Sinanthropus Pekinensis* had lain buried in a limestone hill in northern China. Then, only a short fifteen years after excavation, all remains of Peking Man had vanished. Everything. The tooth that had taken years to come to light and that Davidson Black had cherished enough to make a special capsule in which to carry it; Pei Wenzhong's first *Sinanthropus* skull, which he protected with his life; Jia Lanpo's three skulls, which he had spent three years of his life searching for; plus the scores of bones, altogether representing about 40 individual men, women and children from a stage of evolution previously unknown, and the emotional and intellectual investment of hundreds of scientists and technicians. All had disappeared; lost, stolen or, most likely, smashed to pieces. Gone.

The Chinese scientists at Peking Union Medical College had seen the writing on the wall. As the clouds of war between the Japanese and the United States were gathering in 1941, they sneaked out of Peking in January and made their way to remote Chungking, the bolthole of Chiang Kai-shek's Chinese national capital. They entrusted their loyal technician Jia Lanpo, the discoverer of the three latest skulls of *Sinanthropus*, to take charge of the remaining Chinese personnel at the College. He was told to hold the fort for as long as he could, but to go south if things turned worse.

From Chungking, the Chinese scientists wrote anxious letters about the potential fate of the fossils of *Sinanthropus* should they fall into the hands of the Japanese. The alternatives were either that the fossils be sent to them in Chungking—a long and uncertain journey—or that they be dispatched to the United States. In early 1941, it was decided that, since the Japanese had not yet bothered the American-funded College, the fossils would be most secure there, in a sealed safe.

But the political situation worsened week by week and still no permanent hiding place was found for the treasures. It seemed that foreigners and Chinese alike had become so used to living in Peking with the constant feeling of emergency that they had acquired some sort of immunity to the fear. Otherwise it is hard to explain these months of inactivity.

The fossils had lain buried for hundreds of thousands of years—yet the thought of returning them to Mother Earth for safekeeping does not seem to have occurred to anyone.

In Java, the Dutch palaeontologist Ralph von Koenigswald, who continued Eugene Dubois' search for the roots of humanity in Indonesia, buried priceless fossil skulls in his garden in anticipation of the Japanese invasion. As expected, he was put into prison camp, but—barring one skull—his rich collection of fossils was saved and later, after liberation (from the ground and from the Japanese) went with him to America. Nevertheless, von Koenigswald, like a shepherd who had brought his flock home safely except for one lost lamb, felt the loss of the skull badly. Miraculously, however, a palaeontology student, who became a lieutenant in the war, was assigned to the Military Intelligence Service in Japan in 1945. He wrote to the American Museum of Natural History to ask if there was anything anyone wished him to do while he was in Japan. Somewhat in jest, the reply was to find von Koenigswald's lost skull. Much to everyone's surprise this was followed by an official request from Washington asking for a precise description of the skull. Then, one morning, a few days before Christmas 1945, the lieutenant arrived at the Museum carrying the box with the lost skull. He had been commissioned to bring it back by hand and deliver it to von Koenigswald, who probably had never received so welcome and unexpected a Christmas present in his life. The skull had been found in the Emperor's Household Museum as part of the Imperial collection of curiosities. It seemed that the Emperor had a penchant for fossils.

Weidenreich, in Peking, was not immune to fear. His earlier medical career in Germany had been brilliant and, once he had become interested in the evolution of humanity, he had published an outstanding report on a Neanderthal fossil in 1928. As a Jew, however, he found Nazi Germany intolerable and terrifying and his post in China was his escape. With what must have been a sense of *déjà vu* he fled Peking for the United States in April 1941, taking beautifully prepared casts, photographs and detailed drawings so that he could continue his study of *Sinanthropus* in New York.

Weidenreich was deeply concerned about the safety of the fossils. He tried desperately, before leaving Peking, to persuade the United States ambassador and the commanding officer of the Marine Corps in Peking to send them in official US baggage, thus avoiding the red tape of custom's regulations. He then considered taking the fossils with him in his own personal luggage but decided that the risk was too great. He wrote:

If they were discovered by the customs control in an embarkation or transit port, they could be confiscated. In addition, it had to be taken into account that the objects are too valuable to expose them to an unprotected voyage in so dangerous a time. Considering all the pros and cons we decided, at least for the moment, it would be wise to leave the originals where they are now, that is in the building of the Department of Anatomy at the Peking Union Medical College.

He then asked his technician Hu Chengzhi to make copies (casts) of all the *Sinanthropus* fossils and to send them to him in America. It was a huge job and Hu questioned it. Weidenreich was adamant. 'It's urgent,' he said, 'so get started. The new finds first, then the older ones. If you run out of time, well, just do as much as you can.' Weidenreich also said that the originals should be transferred away from the Japanese-occupied areas and that he would speak to the head of the Chinese Geological Survey to make sure that that happened.

Six months passed while negotiations about the fossils continued between the American ambassador and the head of the Chinese Geological Survey. By then the political situation was irreparable and a final desperate decision regarding the fossils was made as December 1941 approached. The fossils were hurriedly taken from the safe and replaced with a cast of a skullcap. The originals were packed, at night, into two crates by two skilled Chinese technicians, one of them Hu Chengzhi. He wrote of the experience:

> We wrapped every fossil in white tissue paper, cushioned it with cotton and gauze and then over-wrapped them in white sheet paper. The packages were placed in a small wooden box with several layers of corrugated board on all sides for further protection. These boxes were then put into two big unpainted wooden crates, one the size of an office desk, the other slightly smaller. We delivered the two cases to the head of Controller Trevor Bowen's [the American administrator] office, at the Peking Union Medical College, and from then on none of the Chinese knew what happened to them.

Jia Lanpo, meanwhile, had tried to make a run for Chungking. He arrived in Nanking the day after Pearl Harbor and then, discovering that all routes to Chungking were cut off by the Japanese, hurried back to Peking. There, Jia discovered that, in his absence, all United States citizens had been imprisoned and all the buildings of the College were guarded by Japanese occupation troops. Jia still had the proper credentials and was permitted to enter the Cenozoic Research Laboratory, but he knew the Japanese would search him thoroughly when he left. Nevertheless he felt that he had to try to save whatever he could of the precious Chinese dragon bones.

Using impeccable logic, he realised that without the drawings of the Peking Man site, which indicated the positions of the fossils when they were found (and hence their age), the specimens (including casts) would become useless. Duplicates of the large original size drawings and maps would be conspicuous, so Jia scaled-down the drawings by half. Using the thinnest kind of paper, which would pass for toilet tissue, he made rough sketches in the office, which had to be worked over at night in the relative safety of his home. The originals had been his own handiwork, so it was easy for him to make the copies. Still, he couldn't finish more than two drawings a day and it was two months before he got all the drawings copied; as it turned out, just in the nick of time, since the Japanese closed the College and moved troops into the buildings in late January 1942. Jia had also taken most of the photographs showing the work in progress at Chou K'ou Tien, smuggling them out carefully day by day.

In fact, to the Japanese military, the fossils were, if anything, incidental. When, in April 1942, the Japanese military police needed suddenly to occupy a building where boxes of ancient fossils—the full prehistoric record of China's fauna—and books were kept, they ordered the bones and books to be burnt and destroyed. An eye-witness account states that most of the books were scavenged by local residents who later sold them to second-hand book dealers. The bones, however, were scattered and smashed on the ground.

The value of the Chou K'ou Tien fossils was not lost on Japanese academics, however. Pei Wenzhong, by now the head of the Cenozoic Research Laboratory, recalled that 'before the Pearl Harbor attack, Kotondo Hasebe and Fuyuji Takai, both scholars from the Tokyo Imperial University, had come to Beijing. After the Union Medical College was taken over by the Japanese army, they hurried to find the Peking Man fossils. When they ordered the safe opened and saw that there was a copy of the skullcap, they left without a word'. Dr Fortuyn, being a citizen of German-occupied Netherlands, was treated as an internee, rather than a prisoner, and on 9 December, he went to the College as usual. There he found the two Japanese scholars who, strangely, did not ask him about *Sinanthropus pekinensis*. Presumably, having witnessed the generally destructive behaviour of the troops, they accepted the irretrievable loss of the fossils.

Not so the Emperor of Japan, however. When he read the report of the scholars, which had taken several months to filter through to him via the Japanese Ministry of Education, he ordered an intensive search. A series of 'visitors' from Tokyo interrogated Pei and when that proved unsuccessful, a sharp and efficient detective probed into the fossils' disappearance, warning Pei not to leave home so that he could be available for further questioning.

Even Weidenreich's secretary, who had stayed behind in Peking after Weidenreich left, was taken prisoner by the Japanese at that time and asked to search for the two Peking Man fossil boxes in various coastal warehouses. The search was in vain and she was released.

Fortuyn, too, was asked to return to the College, in June 1942, to look for the catalogue of *Sinanthropus* material. But when he went to the College he found it in chaotic disarray. Another Japanese doctor, who had once been on friendly terms with the College staff, interrogated him about the whereabouts of the fossils. Fortuyn, trying to counter the curiosity, said they were partly in Holland and partly in Java. The Japanese doctor denied this saying that Weidenreich had told him they had been sent to the United States five years previously. This rumour had been deliberately spread to protect the fossils. The rumour certainly confused the Japanese who were on their trail, but it didn't save the fossils.

After meticulously packing the fossils in their shock-proof little boxes and then in two crates, Hu Chengzhi delivered them to Administrator Bowen. The next day Bowen transferred the crates to the United States embassy. From there they were transferred to the United States Marine Corps. The plan was that the crates would be shipped onto the *SS President Harrison* along with the marine personnel.

At this point reliable history of their fate comes to an end.

The *President Harrison* was due in the port city of Chingwangtao on 8 December, but she never reached it. En route from Manila, she was pursued by a Japanese warship and ran aground outside the mouth of the Yangtze River. Meanwhile, some of the marines who arrived at Chingwangtao were captured by the Japanese and sent back to Peking. Their personal belongings, possibly including the crates containing the world's greatest archaeological finds, had been sent to the warehouses at Chingwangtao. These warehouses were subsequently ransacked and plundered by Japanese soldiers. The more likely fate, however, was that the cases containing the fossils had been assigned to the last group of marines to be evacuated. They took the train to Tientsin. The train was halted by Japanese troops, who ransacked the luggage, including perhaps the treasure boxes. As a result, the unassuming-looking fossils would have been scattered and smashed along the railway line, the protective tissue paper, cotton and gauze left blowing in the wind.

5
The
supporting
act

Today, 60 years later, the Chou K'ou Tien (now spelled Zoukoudian) sample is still regarded as one of the largest human collections from a single site. And Weidenreich's monographs and papers, written from 1937 to 1948 (when he died prematurely) provide what are still the most detailed descriptions and analyses of any fossil human collection, even though they remain unfinished.

Weidenreich died the year before excavations were resumed at Dragon Bone Hill. In 1949, Jia Lanpo's first return visit there was bitter-sweet. The workers who had survived were overjoyed to see him.

But the fates of the missing workers, such as Tang Liang, a highly skilled fossil hunter who had worked for years on the site and who had died of starvation as a rickshaw coolie, caused deep grief.

Desolate and bleak, also, were the hills surrounding Dragon Bone Hill. The forest that had clothed them had been levelled by the Japanese troops who feared the trees would provide hiding places for guerrilla fighters. Similarly, a temple built in the Ming dynasty on the slopes near Dragon Bone Hill had also been destroyed. It seemed that the dragon spirit had been broken.

The site itself, however, had come through the bitter fighting unscathed, thanks to the foresight of the fleeing workers, who, in 1937, had thrown a thick layer of rubble over it as protection. They must have used unseized and unsorted material from layers already dug because three *Sinanthropus* teeth were found in the rubble as it was removed. It was a hopeful sign but, unfortunately, resources for the dig were few and politics in China continued to play havoc with palaeontology with the onset of the 'cultural revolution'.

It was 1978 before the Peking Man site once again became a project of the Chinese Academy of

Sciences. But by then the work was no longer ground-breaking and the site had been eclipsed by major finds elsewhere on the globe, particularly in Africa, where Charles Darwin had, all along, predicted our roots to be.

A warm, swampy place with tall trees and meandering streams is a predictable place for roots. Not so predictable was the location—the Fayum region of Egypt, near where Cairo is now situated. Thirty five million years ago, during the Oligocene epoch, this was the tropical womb of humanity. In it lived a variety of fabulous-looking creatures, including dozens of different types of primates, a higher number than found in even more varied environments of today. One of these creatures, a nondescript, small-brained, cat-sized mammal, was our distant ancestor.

It was a fruit-eater, crawling on all-fours above the branches and suspending itself by its powerful hind-feet to eat. While the brain was small it had an ape-like organisation emphasising a large visual area. Its name was *Aegyptopithecus* (the Egyptian ape). It begat (in an evolutionary sense), among other things, *Afropithecus* (the African ape), which is the common ancestor of the Old World monkeys, apes and, ultimately, hominids.

Afropithecus was the size of a large dog, but one that bounded about on branches. It lived 20 million years ago and had a robustly built skull with large jaws, and teeth with thick enamel. Its food items clearly required grinding and crushing. It inhabited a drier forest than *Aegyptopithecus* and occupied a dietary niche of hard-to-peel fruits and other hard reproductive parts of plants, such as seeds and other fibrous items which required more muscle forces to chew.

About 15 million years ago, the central highlands of Africa were uplifted as a result of continental drift between Africa and Eurasia and more open-country habitats appeared with open woodland and bush environments. East Africa continued to dry and the environment diversified into mixed habitats with grasslands, bushlands, as well as the jungles in lower areas. It was only at this time that animals which we usually associate with the grasslands of Africa, such as giraffe and antelope, first appeared co-evolving with the savanna.

The changes in environment also catalysed evolution among the primates, and a group known as the Dryopithecines emerged from their *Afropithecus* ancestors. Dryopithecines were a spectacularly successful group of apes diversifying into a variety of forms; from small and graceful to large and bulky, similar to orang-utans; from strictly tree-dwelling species to species with mixed terrestrial and arboreal adaptations.

Among the many species of Dryopithecines was the last common ancestor of the living hominoid primates—gibbons and great apes—including chimpanzees and humans.

Dryopithecines were the first apes to migrate across the Old World, and in doing so they gave rise to distinct geographic evolutionary lines: the apes of Eurasia, of which only the orang-utan remains, and the Anthropithecines—the 'man-apes'—of Africa.

The Anthropithecines had thick enamel teeth and broad, deep cheeks. Perhaps 30 kilograms in weight, they had a mixed tree and ground-dwelling adaptation. In anatomy, diet and habitat, comparison with any species now living is difficult. Superficially, Anthropithecines looked like gorillas in facial appearance (in the same broad sense as we do). They had longer legs and shorter arms than chimpanzees. Their hands, which cannot be compared with any thing living, were enormous. Anthropithecines were semi-terrestrial and showed a tendency to use their legs for walking bipedally.

All except two lines of Anthropithecines eventually became extinct in response to a general drying of the environment, and an increase in tough, nutritionally poor and generally low quality vegetation.

The frugivorous Anthropithecines, with large jaws and teeth to cope with anything tough, had plenty of opportunities to diversify into the evolving patchwork of microhabitats. They could scramble about on the ground and they could climb in the trees. About nine million years ago, one group specialised by increasing enormously in body size and focusing on ground-layer foliage. This group eventually evolved into the (now mostly extinct) gorillas.

Not so fussy were another line of Anthropithecines who evolved a more eclectic diet, including meat, but still making low quality food an important part of the diet for at least part of the year. This group focused on cooperative hunting, food-sharing between adults of both sexes and using crude tools (such as a stick) in foraging. Socially complex, this group was physically diminutive with large, thick enamel teeth and reduced canines. About six million years ago these Anthropithecines split into a line that developed below-branch feeding (that is hanging from one arm while the other grabs the food), as well as a somewhat inefficient way of walking— on the knuckles—which, however, had the advantage of freeing one arm for carrying. The last surviving species of this line is the chimpanzee.

The other line, the Australopithecines, took a different tack and focused on keeping their two feet on the ground as walkers. Their environment was primarily wooded or forested, with occasional forays into the expanding savanna. While they walked bipedally, they still maintained powerful forearms for climbing. In size, even the bigger males (often twice the size of females) were smaller than today's Pygmies.

The teeth of this early Australopithecine reveal a low quality diet that required processing by chewing. Just as well, since the hands, while developing a longer thumb, still lacked the ability to grip in order to make tools to help process poor quality food. The powerfully developed masticatory apparatus resulted in large faces and jaws since large bones were required as superstructure on which to attach the enormous muscles of the jaw. Their brains, however, were small, about the size of a chimpanzee's.

Walking on two legs rather than four, opened the possibility of a more complex feeding strategy based on cooperation, and a sexual division of foraging that broadened the potential food resource. Carrying

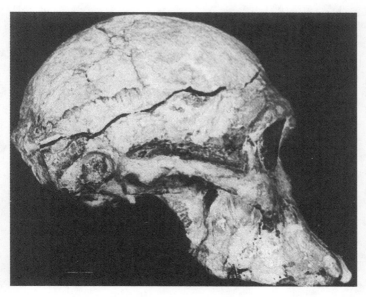

Australopithecus africanus.

became important for weapons such as clubs, and tools such as digging sticks.

Australopithecines were the earliest hominids and represent the longest lasting era—about four million years—pre-stone tool-making hominid forms. If ever there was a missing link between apes and humans it was to be found amongst the Australopithecines.

Ernst Haeckel, the nineteenth-century scientist and disciple of Darwin, who saw the study of fossils as superfluous in the debate about the descent of man, nevertheless accurately envisioned an Australopithecine-like creature as our ancestor 50 years before the first one was discovered by Raymond Dart in 1924—five years before the Peking Man was found.

Raymond Arthur Dart was born on 4 February 1893 in Queensland, Australia. Raised on a cattle station, the fifth of nine children, he excelled at school, won a scholarship to the University of Queensland and then continued to study medicine at Sydney. After serving in the Medical Corps in the First World War, he studied in England under

Grafton Elliot Smith, the same neuroanatomist who earlier took Davidson Black under his wing.

In 1922, Dart accepted the inaugural seat of anatomy at the newly-founded University of Witwatersrand in Johannesburg, South Africa. It was here, in South Africa, that Dart made the discovery of a fossilised skullcap and jaw of a very young child in two- to three-million-year-old lime debris left by a mine excavation into the remains of a cave near Taung.

Dart considered the fossil the missing link between apes and humans and named it *Australopithecus africanus*, the man-ape of Africa. The 'Taung child', as it came to be known, had a number of outstanding human features. In the first place it had small, flat-wearing canines, like ours. Second, the pattern of the brain which was imprinted on the inside of the skull suggested that the brain, although small, was slightly larger and structurally different from that of an ape. Third, the position of the foramen magnum—the hole in the skull through which the spinal cord passes—was at the bottom of the skull (in apes it is towards the back of the skull, because of their crouching posture). The position of the foramen magnum in the Taung child indicated an erect posture with the head balanced on top of a vertical hominoid spine.

These three features were precisely what Darwin had predicted for hominid origins—reduced canines, bipedalism and changes in the structure of the brain.

Predictably, with Piltdown Man still in vogue which, with its large human brain and ape's canine, had the opposite features of the Taung child, and Eugene Dubois still smarting after suggesting that *Pithecanthropus erectus* might have been a human ancestor, most scholars at the time interpreted the Taung fossil as an early ape. Dart ignored them. As far as he was concerned here was a species of tool-using, hunting hominids that employed a combination of intelligence, brute strength and weapons. Dart finally travelled to London in 1930 to try and win support for his Taung child, but his finding was overshadowed by the recently discovered Peking Man.

It is ironic, to say the least, that Dart had indeed discovered a human ancestor millions of years older than *Sinanthropus*, and unearthed even before Davidson Black had excitedly described his first *Sinanthropus* tooth, going out on a limb to state that it was a relic of a new type of early man. In reality, Peking

Man was just the supporting act for Dart's main feature.

Perhaps the most famous Australopithecine is 'Lucy'. It took 17 more years and several more discoveries of Australopithecines in the same area, by different scientists, before the Taung child was finally recognised as a crucial missing link. A little over three million years ago, earlier than even the Taung child, a diminutive Australopithecine female lived by a lake on the edge of the lush forests in the Great Rift Valley of East Africa. Dying a natural death, her body simply sank into the soft sediments of the lake, the flesh slowly rotting off the body. Heavy rains washed sand over her, and over millennia hundreds of feet of sediment built up, burying the bones deeper and deeper. Molecule by molecule her bones were turned to stone.

Later, movements of the earth's crust ripped open the rift even further and gradually brought her ancient grave closer to the surface. Rains cut down through the now dry earth of Ethiopia until one heavy storm washed her bones clean of sand, bringing them into the sunlight once more. Harder to find, and more precious than diamonds, it was anthropologist Donald Johanson who stumbled over them.

In 1973 Johanson was walking the desert ravines and valleys of the Great Rift Valley near a place called Hadar, in Ethiopia. Here, the ravines and valleys had cut through ancient sediment out of which fossils were emerging. Luckily for him, the seasonal rain did most of the hard work, scouring the barren earth and uncovering buried fossils. One afternoon, while collecting elephant teeth, he looked down and found a couple of pieces which joined to form a familiar looking knee-joint. He tucked the fossils away and brought them back to America where he showed them to an anatomist. They were indistinguishable from the joints that make up a human knee. Johanson had stumbled across the remains of the oldest human ever found.

The next year, the team of anthropologists was back at Hadar. It was about noon on a searing day when the diminutive female was found. Johanson was walking back to his Land Rover and his glance happened to fall on a bone resting on the surface of the ground. He scanned the slope and could see bits of a leg, a skull and fragments of a jaw. He had stumbled across what was most of an entire hominid skeleton, although it was partially scattered perhaps by the cattle of the nomadic Afar people.

The precious fragments were taken back to camp, and the team of anthropologists worked into the night piecing the puzzle of bones together. A Beatles song was playing in the camp: 'Lucy in the Sky with Diamonds'. Out of the puzzle emerged a tiny adult female. Lucy.

Lucy is the earliest female hominid to leave her dentition and remarkably complete skeleton to posterity. At only 105 centimetres (just under 3.5 feet) in height, and weighing about 27 kilograms, Lucy was a midget. Males of her species *Australopithecus afarensis*, while still the size of Pygmies, were much larger than her. Despite her size, she was muscularly powerful as a result of constant exercise. A veritable iron-woman, her muscular condition and the evident strength in her forearms indicate that climbing was still part of her lifestyle, perhaps to find a safe place to sleep or to steal a carcass left by a leopard. As the forests continued to shrink, Lucy and her kind walked across the grasslands to reach the clumps of trees where her main food was found. Her hands were free to collect and carry the valuable food she found. This slight advantage was all she needed. While other apes declined and became extinct, the Australopithecines flourished.

With a receding forehead and a braincase not much larger than a chimpanzee's, Lucy was no smart ape. Nevertheless, her brain was undergoing a major neural reorganisation that involved an expansion of those areas associated with cognition, categorisation, symbolisation, determining relevant from irrelevant elements and speech. Neural reorganisation made certain kinds of complex behaviours, such as mentally mapping the position of resources across the environment, much easier to learn for our ancient ancestors.

Lucy had a large flaring pelvis. These good child-bearing hips allowed room for the strong, wide shoulders of the baby, the head being rather diminutive, due to its smaller brain, compared with ours. She could give birth, if not with ease, then at least without obstetric difficulties. The shape of the birth canal, however, indicates that the birthing process was a slow one, probably laborious and dangerous enough to be a social rather than solitary event.

Her hands show a small but powerful individual with flexible fingers and a wrist that could easily manoeuvre. While the fingers were long, about the same size as ours, they were not flattened at their tips as ours are. Her thumb was shorter than ours too so that she had difficulty using the precision grip, that is, gripping with the thumb and one or more fingers, with the palm being used as a passive prop—the sort of grip we use every day. With such hands, she could not make tools.

Importantly, her hands show no evidence of knuckle-walking. In other words her species was fully bipedal and walked like humans 3.5 million years ago. She had shorter legs, however, than ours. Her legs show that she could run swiftly for short distances, even though she was not so efficient at walking long distances. Her foot was human-like, with an arch and no adaptations at all for grasping and fully adapted to walking on two legs. Virtually every modification in the skeleton for walking bipedally, found in humans, is found in Lucy.

Almost as if to prove the point, an exciting discovery of ancient hominid footprints was uncovered in Tanzania at about the same time as Lucy was unearthed in Ethiopia. Two parallel trails, containing a total of 54 clearly hominid footprints—so good that they allow clear recognition of soft-tissue anatomical features such as a heel, arch and big toe—were excavated in 1978 and 1979 at Laetoli.

Laetoli lies in the eastern branch of the Great Rift Valley of East Africa, a tectonically active area. About 3.6 million years ago a volcano, located about 20 kilometres away from Laetoli, began to belch out clouds of ash which settled in layers on the surrounding savanna. At one point in the volcano's active phase, a series of eruptions coincided with the end of an African dry season. After a light rainfall, the animals that lived in the area left their tracks in the moist ash. The material ejected from the volcano was rich in a mineral called carbonatite (which acts like cement) so that when hardened, it set the thousands of footprints that covered the area. Shortly afterwards the volcano erupted again, burying the footprints and preserving them.

The Laetoli footprints show two side-by-side trails made by two individuals walking bipedally: the smaller footprint presumably a female, walking with a male. The footprints show a pronounced heel strike, a big toe in front of the ball of the foot and parallel to the other toes and a deep impression for the big toe showing that the creature walked with toe-off—like us. The stride length is very short. They were walking slowly, probably strolling.

The footprints fit Lucy's feet like a slipper.

While Lucy had delicate Cinderella-feet, the fairy-tale line that most suits her is 'Grandma, what big teeth you have'. Lucy, while indeed likely to have been our grandmother (in species terms), had teeth 3 times larger than ours. This resulted in a very large protruding face, which reflected numerous adaptations to powerful chewing such as widely flaring, well-developed cheeks to support the muscles of the jaw and a flattened nose.

Mammals can have four distinct types of teeth (usually not all together)—incisors, canines, premolars and molars. Canines and incisors cut and shear. Molars grind and crush. Humans are the exception because we have all four types. This in turn is a reflection of our adaptability in diet—we are omnivores which means we eat animals' parts, as well as the reproductive parts of plants (fruit, flowers, buds, seeds) and the structural parts (leaves, stem, bark). It is also a reflection of our evolutionary roots.

Our distant ancestors were frugivores—species relying on fruits, and occasionally nuts and seeds, in their diet. Frugivore species tend to have large, protruding front teeth to open nuts, seeds or fruits, and low, flattened back teeth for grinding and crushing. Various primate species have adapted diets that emphasise nuts and seeds and require a great deal of force during chewing, thus evolving relatively large, ape-like back teeth (molars) with thick enamel.

Even our earliest ancestor, the cat-sized *Aegyptopithecus*, had ape-like molars with thick enamel and spade-like incisors. Molars were part of a powerful masticatory apparatus that resulted in powerful jaws and muscles. These were firmly anchored in the skull, which formed a crest for its entire length as a structural adjustment to the big teeth, suggesting a diet of tough leaves and other structural parts of the plants, as well as fruit.

Afropithecus was a much larger primate than its ancestor. It was robustly built with the regulation large jaws and teeth. It had an adaptation to dietary items requiring both grinding and crushing. *Afropithecus* lived in slightly drier and more open woodland environments than *Aegyptopithecus*. Its large, buttressed teeth suggest a diet of harder-to-peel fruit and other hard reproductive parts such as seeds, as well as a more folivorous element (for example, leaves) with plenty of fibre. *Afropithecus* had a greater bite force than *Aegyptopithecus*. It was these characteristics—the ability to bite and chew (rather than to cut and slice)—that set us up for success.

Lucy's teeth were larger than a 200 kilogram gorilla's. Her large front teeth were probably used in the same way that a gorilla places a large leaf in its mouth and pulls the stem out between semi-clamped front teeth in order to strip off the edible cellulose. Lucy's teeth indicate a broad variety of food, probably seasonally varied, with emphasis on low quality food for at least part of the year. She was also an active scavenger of meat. Lucy's teeth had a longer effective life, because of their thick enamel. It meant she could eat more tough food. Even so, at 35 Lucy would have been an old lady with her teeth worn flat.

It is important to understand this emphasis on teeth. The saying 'we are what we eat' could not have a more profound meaning than for humans. This is simply because it was what we ate that ultimately dictated what we became. The ability to concentrate on a broad variety of poor quality food meant that our ancestors could exploit a diversity of habitats. This ultimately led us down from the trees on two legs, which in turn provided more opportunities (as long as you could eat tough food) and freed the hands for carrying and for exploratory use of make-shift tools such as sticks for digging out roots. No doubt an ability to mentally map the position of resources across the environment paralleled our ability to move across different habitats, triggering a reorganisation of the brain, which resulted in a swelling of those lobes relating to speech and culture, at the expense of those relating to vision and smell.

About 2.6 million years ago there were at least two contemporary species of Australopithecines, in species terms the children of Lucy: *Australopithecus africanus* (the Taung child) of South Africa and *Australopithecus aethiopicus* of East Africa. Someone at the time started banging rocks together and made an unprecedented and revolutionary discovery—the

technology of making stone tools. This discovery would, like the butterfly wings of chaos, reverberate down through time affecting the existence of every species, not just hominids. The identity of the discoverer is lost in the mists of time, but the most likely candidates are probably all the hominids because the saying 'monkey-see, monkey-do' is as apt for hominids as any monkey.

Essentially, the first tools were crude and opportunistic. The maker showed little understanding of the mechanics of stone fracture, striking rock against rock in a haphazard manner, afterwards carefully picking through the pile of flakes for one or two that would be most useful for a specific purpose. Perhaps a chopper for cutting wood, or a long, thin flake that could be used to cut cartilage and tendon.

Despite the opportunistic fashion of the tool making, the fact that the maker could associate the form of a flake with a specific function was a milestone in human evolution. Stone tools, however, represent only a small part of the tool-making activity of early hominids. Because they are well preserved we know much more about stone tools than any made of wood or bone. Simple digging sticks, for instance, made of bone or antler opened the possibility of a vast source of edible roots that could not be dug up by unaided hominid hands. And in the drying, seasonal climates of Africa this would have provided an edge in the struggle for survival. Interestingly, hundreds of thousands of years lie between the first appearance of artefacts and the first appearance of any *Homo* species.

About 1.9 million years ago, a *Homo*-like species— *Homo habilis*—had evolved, probably from *Australopithecus africanus*. It shared its era with several other species: *Australopithecus robustus*, *Australopithecus boisei*, and a possible, as yet unnamed new species. *Homo habilis* was initially discovered by the Leakey family in Olduvai Gorge, Tanzania, in the early 1960s. Louis Leakey named it *Homo habilis* ('handy-man', because it was found with many tools) in 1974 amid much controversy. He claimed it was our direct ancestor. Because the Leakey family were funded and well publicised by the *National Geographic* magazine, they and their fossil became famous.

Today controversy still rages about the little handy-man. Mainly this is because, like all the other Australopithecines, it is a midget and therefore difficult to differentiate from them. Part of the problem is that the fossils attributed to handy-man are so diverse. Some share common features with *Homo sapiens* and some with the Australopithecines. There has even been the suggestion that *Homo habilis* simply represents a grab-bag of assorted fossils from the latest Pliocene to early Pleistocene. As for the tools found with the original Leakey discovery, there is no evidence to suggest any advance on the activities of the other Australopithecines.

One thing is certain though, *Homo habilis* cannot have been our direct ancestor because *Homo sapiens* was there first.

The first evidence of the new kid on the block is at 2.2 million years ago—much older than the oldest handy-man fossil. That this was a totally new creature is unequivocal. It was twice the height and more than double the weight of the local Australopithecines. *Homo sapiens*—a human being—had arrived.

Who was its direct ancestor? Certainly one of the Australopithecines. They were also evolving in the same direction, that is, larger brain size, reduction in tooth size and increased use of low-quality foods through access to tools. But which Australopithecine? None of the known Australopithecines quite fit the picture. Rarer than diamonds, mother earth still guards her secrets like a dragon its jewels.

The highly successful Australopithecines, who had been around for five million years, no doubt kept the newcomer on its toes. Indeed the East African *Australopithecus boisei* kept up with humans for the better part of three-quarters of a million years. Competition for at least some limiting resources between tool-using, intelligent hominids could be expected to have been intense. With the Australopithecines forcing the evolutionary pace, humans had to do something else. Their solution was not only to use tools as part of the basal hominid strategy—monkey-see, monkey do—but to make them part of their culture.

Up until about two million years ago, tools were little more than an adaptive response to certain seasonal requirements and were used, for example, to dig up a tuber, and then abandoned. In humans, tool use evolved as part of a cultural system and became incorporated within their behavioural repertoire. Rather than be manipulated, humans began to manipulate.

Paralleling the migration out of Africa was the development of a special-purpose tool kit. Humans had evolved the ability not only to match form with function but to create a preconceived form out of formless stone. Materials for tools were well chosen and people understood the flaking properties of specific rock types. Tool concentrations—sharpened flakes and a few large cutting instruments—began to appear at sites where an animal was butchered. The faunal remains indicate that humans regularly consumed meat from medium- and large-sized mammals, including giant baboons. This strongly suggests hunting and confrontational scavenging was a cooperative and organised activity, probably dependent on networking between family groups or indicating seasonally flexible populations.

The writing was on the stone for the Australopithecines and they became extinct around 1.4 million years ago, at about the same time as humans began their migration out of Africa, leaving them as the only hominid on the planet for the first time.

6

Whatever happened to *homo erectus?*

After analysing Peking Man and Java Man fossils Franz Weidenreich argued, in 1943, that only a single genus was necessary to describe the Pleistocene hominids. He suggested that *Pithecanthropus erectus* and *Sinanthropus Pekinensis* be called *Homo erectus*. In this way the evolutionary process would not be clouded by the taxonomy that was obscuring it. Once done, it was easy to see that Peking Man and Java Man were the same species; a big bodied, large brained, high-energy requiring animal where foraging was based on the sharing of foods obtained by hunting, scavenging and gathering.

Homo erectus had a distinct origin around two million years ago, clearly splitting off from the Australopithecine lineage. It was able to successfully colonise much of the world including remote areas of Indonesia and China by half a million years after its appearance.

Like most species, *Homo erectus* was not a static species. About a million years ago it had reduced back teeth and larger front teeth. This was a reflection of the species' evolving tool kit. Technology allowed the processing of tough food and the reduction of the back teeth, but the front teeth became used as a third hand to grip onto something while manipulating with the hands. Different structural characteristics are associated with this activity which place more pressure on the brow ridge—hence the 'veranda' above early Pleistocene faces.

Further technological innovations and refinements in the middle Stone Age (around 750 000 years ago) resulted in more efficient tools such as knives, burins, drills, scrapers and even composite tools, which in turn led to a more effective application of leverage and force. There were significant decreases in the use of the jaws to hold and pull— the vice-like actions that had been so important earlier. A series of biomechanical changes took place including the snout reducing to a jaw and the region on either side of the nose receding, resulting in the nose having more profile. The brow ridge, which acted as a buttress, eventually disappeared.

The result was an animal that looked like *Homo sapiens*.

Franz Weidenreich, too, noticed that certain characteristics of *Homo erectus*, such as flatness of the face and flatness of the nasal saddle—the top of the nose—carried through to *Homo sapiens* populations of mongoloids. More recently, Australian Alan Thorne also has found a sequence of skulls, starting in Java a million years ago, which link prominent brow ridges, foreheads and faces from *Homo erectus* to similar characteristics of *Homo sapiens* in Australia. He concludes that there is regional continuity. Others, however, insist that these are two species.

But the problem with this is that regional continuity across a speciation event (from one species to another involving different features in different regions) is simply impossible. It would be like nominally naming a brand new species of *Homo* from the year 2000, let's call it *Homo i.t.* In the Kalahari, *Homo i.t.* has populations that are diminutive and brown;

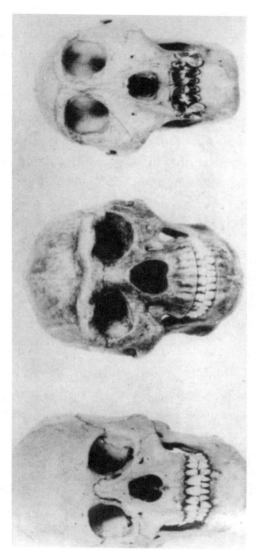

Female *sinanthropus* (centre) compared with a female gorilla (right) and a modern Chinese (left).

populations in Ethiopia that are tall and rangy; populations in northern Europe that are solid and white; populations in Asia that have flattened faces and small noses and so on. Confused? Of course. That's because the difference between *Homo i.t.* and *Homo sapiens* is no more than a line drawn in the loess. With the same regional differences, occurring across the alleged speciation event, they must be one and the same species, that is, *Homo sapiens*.

The price paid for excessive taxonomising is confusion.

Weidenreich went further than suggesting that only a single genus was necessary to describe the Pleistocene hominids. Following his logic through to its logical conclusion, he was the first to suggest that there was good reason to sink *Homo erectus* into *Homo sapiens*.

His explanation was this:

If the *Hominidae* are one species in the genetical sense and an exchange of genes was possible throughout their evolution...the commonly used form to represent their lineage gives an entirely wrong idea. The tree with a common stem and more or less abundant ramifications leaves no possibility to indicate graphically an exchange of genes. The branches and sub-branches appear to evolve completely independently of each other once they have deviated. In reality, there must have been intercommunication between the branches. The graph which best fits this perception is a network. Its interconnections indicate the lines along which the exchange of genes could be effectuated.

In other words a braided stream of evolving populations, each adapting to local conditions, yet all consistently linked by gene exchange.

Today there is an increasing movement towards Weidenreich's way of thinking. Largely this is because there is nothing to mark the beginning of *Homo sapiens*. No single definition has been found that distinguishes *Homo sapiens* from *Homo erectus*. Species are similar to individuals in that they have real beginnings and endings with their own evolutionary tendencies in between. Neither *Homo erectus* nor *Homo sapiens* alone fit this description, but a lineage combining the two does.

We are all creatures of our time, and in our time we still insist on looking at ourselves outside the natural order of things. Biologists who study, say, butterflies know that different populations of the same species can look quite distinct. They are not tempted to hastily split them into separate species. We make far too much of our anatomical differences as our fixation on trivial race differences tragically demonstrates. Why must we look at ourselves any differently?

The fact is Peking Man was us.

We shall not cease from exploration
And the end of all our exploring
Will be to arrive where we started
And know the place for the first time.

T.S. Eliot, 'Little Gidding',
from *The Four Quartets*

References and acknowledgments

The book has been written with readability in mind. I have therefore not cluttered the pages with references or numbers. There are a number of books and papers, however, which anchor this story. They are:

Andersson, J. Gunnar (1934). *Children of the Yellow Earth*. MIT Press. Cambridge, Massachusetts.

Dreyer, Edward, L. (1995). *China at War, 1901–1949*. Addison Wesley Longman. London and New York.

Hood, Dora (1964). *Davidson Black: A Biography*. University of Toronto Press, Toronto.

Jia, Lanpo and Weiwan, Huang (1990). *The Story of Peking Man: From Archaeology to Mystery*. Foreign Languages Press, Beijing and Oxford University Press, Oxford.

Ren, M., Liu, Z., Jin, J., Deng, X., Wang, F., Peng, B., Wang, Z. and Wang, Z. (1980). 'Evolution of limestone caves in relation to the life of early man at Zhoukoudian, Beijing', *Scientia Sinica*, vol. XXIV No. 6.

Sigmon, Becky A. and Cybulski, Jerome S. (1976). *Homo erectus: Papers in Honor of Davidson Black*. University of Toronto Press, Toronto.

Theunissen, Bert (1985). *Eugene Dubois and the Ape-Man from Java: The History of the First 'Missing Link' and its Discoverer*. Kluwer Academic Publishers, Dordrecht.

Wolpoff, Milford H. (1999). *Paleoanthropology* (second edition). McGraw-Hill, New York.

I also commend the Internet.

My heartfelt thanks, as usual, to my husband Noel Preece, who took the brunt of the first drafts, and my sweet son Luke Preece for his patience. Friend and mentor Peter Mitchell suggested the topic. Colin Groves put me on the right track with respect with up-to-date palaeoanthropology material. Siobhan Denniss helped with searches for other material. George Moss helped with dragons. The interpretation I have provided, however, is completely my own.

Index